BARBARA J. SIVERTSEN has been managing editor of the *Journal of Geology* for over twenty-five years. She is the author of *Turtles, Wolves, and Bears: A Mohawk Family History* and coauthor of *The Legend of Cushetunk: The Nathan Skinner Manuscript and the Early History of Cochecton*.

The Parting
of the Sea

〜

The Parting of the Sea

How Volcanoes,
Earthquakes,
and Plagues
Shaped the Story
of Exodus

BARBARA J. SIVERTSEN

PRINCETON UNIVERSITY PRESS
PRINCETON AND OXFORD

Copyright © 2009 by Princeton University Press

Published by Princeton University Press, 41 William Street, Princeton, New Jersey 08540

In the United Kingdom: Princeton University Press, 6 Oxford Street, Woodstock, Oxfordshire OX20 1TW

Library of Congress Cataloging-in-Publication Data

Sivertsen, Barbara J., 1949–
   The parting of the sea : how volcanoes, earthquakes, and plagues shaped the story of Exodus / Barbara J. Sivertsen.
       p.   cm.
   Includes bibliographical references and index.
       ISBN 978-0-691-13770-4 (hardcover : alk. paper)   1. Bible and geology.   2. Bible stories, English—O.T. Exodus.   3. Exodus, The.   I. Title.
   BS657.S58   2009
   222'.12095—dc22   2008032722

British Library Cataloging-in-Publication Data is available

This book has been composed in Minion family

Printed on acid-free paper. ∞

press.princeton.edu

Printed in the United States of America

1   3   5   7   9   10   8   6   4   2

In memory of my father,
HOWARD VOLMER SIVERTSEN

Contents

Figures and Tables

〜

Acknowledgments

I would like, first, to express my profound thanks to several very helpful and supportive reviewers, Elizabeth Wayland and Paul T. Barber, Jelles de Boer, and Adrienne Mayor, for their encouragement and many useful comments, and to my editor, Rob Tempio, for his unswerving support, without which this book would not have been published. I also would like to thank several helpful geologists, particularly Alfred T. Anderson, Jr., for answering many questions on volcanology, for reading my chapter on Santorini, and for lending me an eruption video of Mount Kilouea that introduced me to fire fountains and hissing steam vents. Gerald Friedman shared many of his publications with me and offered useful suggestions. Bill Rose introduced me to secondary maxima of tephra falls and to the massive mid-continent tephra deposits that originated in Idaho. Colin Wilson suggested sources of information on Aegean volcanoes, and Jörg Keller and Sharon R. Allen answered my queries with thoughtful letters. I would also like to thank Professor Ioannis Liritzis of the University of the Aegean for sharing information on his finds on Yali with me.

Andrea Twiss-Brooks, the geological sciences librarian at the Crerar library, provided help in getting key references. Christopher Winters, the map librarian at the University of Chicago, found everything from the most modern U.S. Geological Survey data bases and maps to old British Ordnance Survey maps of the Nile Delta. Taeko Jane Takahashi, librarian at the Hawaii Volcano Observatory, found an appropriate picture of a fire fountain for me. I would also like to acknowledge the late University of Chicago and Field Museum Egyptologist Frank Yurco, whose class on the Egyptian background to the Exodus convinced me that the Exodus had an undeniably New Kingdom component; also T. H. and P. H., whose class on the Ten Commandments made me realize how closely the Passover resembled a covenant meal; and Karl W. Butzer, who first introduced me to the study of the environment and its impact on past human populations. Finally, I would like to thank my husband for his unfailing encouragement and support, and for his exhaustive Internet searches for obscure items when I needed them.

The Exodus, Oral Tradition, and Natural History

The story of the Exodus is one of the best known narratives of Western Civilization. As recounted in the Bible, the Israelites are slaves in ancient Egypt. Moses, an Israelite raised in the Egyptian court as the adopted son of Pharaoh's daughter, kills an Egyptian who is mistreating an Israelite slave and is forced to flee the country. He arrives in the land of Midian, meets the daughters of the Midianite priest Jethro, marries one of them, and produces two sons. One day, while tending sheep for his father-in-law on the west side (or the back side, or the far side, depending on how the Hebrew is translated) of the wilderness or the desert, Moses sees a burning bush. The odd thing is that the flames do not consume the bush, and out of it an angel of God speaks. This is the prelude to a series of conversations between Moses and God. God, or Yahweh, wants Moses to return to Egypt and bring Yahweh's people back to the land promised to their forefather Abraham—the land of Canaan.

Moses is more than a little reluctant to take up the task, but eventually he returns to Egypt. Moses and his brother Aaron go to Pharaoh and demand that Pharaoh let the Israelites go on a three days' journey during which they are to make sacrifices to their God. The two brothers perform a series of supernatural tricks before Pharaoh to try to convince him to do what they ask. When these tricks prove ineffective, God inflicts a series of plagues on the Egyptians: the water of the Nile is turned to blood, there are plagues of frogs, gnats, and flies, cattle become diseased, people develop boils, there are plagues of hail, locusts, and darkness. Finally, because Pharaoh refuses to let the Israelites go, God declares that He will kill the firstborn of Egypt. God tells Moses how to arrange for the Israelites to avoid this fate by killing a young goat or sheep, roasting it, and smearing its blood on the doorways of all the Israelite households. After the first-born of Egypt die, Pharaoh tells Moses and his people to leave during the night. They depart immediately, guided by a pillar of cloud by day and fire by night. Pharaoh changes his mind, however, and the Israelites are pursued to the edge of the sea. God saves them by having the waters part, allowing the Israelites to pass through on dry land. After the Israelites have passed through, the waters return, drowning the pursuing Egyptians.

What to make of all this? Is the story of the Exodus real? Did the ancestors of the Israelites really leave Egypt following plague and disaster, cross a body of water that miraculously separated before them but drowned their pursuers, and go to a distant mountain to see God in a cloud of fire and smoke? Did they then wander in the wilderness for forty years and finally cross the Jordan River and conquer Jericho when its walls fell to the sound of their trumpets?

Today, only the most conservative biblical scholars champion a literal reading of the Exodus narrative. The majority of scholars and general readers alike discount such wondrous happenings as the figments of primitive imaginations. Their purpose is theological, their historical value is limited at best. In fact, in the last twenty-five years a group of biblical scholars known as Revisionists or Minimalists has gone so far as to suggest that the history in the Hebrew Bible was an invention of theologically minded writers only a few centuries before the Common Era. Although their opinions are not shared by the majority of biblical scholars and archaeologists, no one has been able to point to any direct textual or archaeological evidence for the historical veracity of the stories in Exodus and Numbers. No archaeological traces can be attributed to the early Israelites in Canaan before the early Iron Age (after 1200 B.C.E.), and there is no evidence of a distinct population of early Israelites in Egypt. The area west of the Jordan River reveals an archaeological picture quite at odds with the biblical accounts of the Israelite journey to the Promised Land, and there is little or no evidence of the Conquest, as it is described in the book of Joshua, in the archaeological record of the Late Bronze Age (1550–1200 B.C.E.) Canaan. In short, in ways large and small, the biblical story of the Exodus, the sojourn in the wilderness, and the conquest of Canaan does not agree with the archaeological picture that has emerged in the past forty-five years.

The Exodus, if it has any historical reality, must have occurred no later than the thirteenth century B.C.E., for toward the end of that century the famous stela of the Egyptian pharaoh Merneptah mentions an entity "Israel" already in existence in Canaan. Yet writing, by most estimates, didn't really get started in ancient Israel until the tenth century B.C.E., or even later. This means that any accounts of the Exodus would have been carried down orally for hundreds of years.

Biblical scholars who study the ancient texts are experts on the languages of the scriptures and on textual and literary criticism. They are, above all, textually minded. To them, the most important part of the texts are their literary elements, as conveyed by *the words themselves*: their meanings, their grammar, their syntax. But ancient Israelite society was overwhelmingly oral, and in oral societies, words are important only insofar as they convey *the story*—the events taking place, not the literary text. Words usually change with each rendition, each oral performance, of a traditional story in a nonliterate society.

I believe that at their basic level, the *stories*—the narrations of the events— of the Exodus, the sojourn in the wilderness, and the Conquest must be seriously considered as oral traditions that may contain remnants of oral history. To do this, to consider the biblical stories as residually oral texts, we must ask several crucial questions: What, exactly, is oral history? What is it capable of transmitting, of remembering? Where and how does it fail?

WHAT IS ORAL HISTORY?

First, oral history is not "history" as we in the modern world have come to understand it. It is more intimate, concerning itself with what happened to an individual or a family or a small group of people. It scarcely ever deals with the great events that define what we call history. It starts with eyewitness testimony, like seventeen-year-old Ensign William Leeke's account of the Battle of Waterloo in 1815.[1] Reading his story, we learn of his dismal night spent in the open before the battle, the rain pelting down, sharing a blanket and some straw with a fellow officer, and trying to avoid two galloping horses. In the battle itself we hear of his carrying of the regimental flag, are told of the peculiar smell of gunpowder mixed with rye, his tears at the sight of the first two dead, and his horror as hundreds more fell. One of the most important parts of the battle for Ensign Leeke was when he failed to draw his sword because the loop of his sword knot had become entangled with the scabbard. Only at the end of the day, when Ensign Leeke's regiment, the 52nd Foot, marched in pursuit of Napoleon's retreating Imperial Guard, do we come into contact with the conventional history of the battle. If this were the only surviving account of the Battle of Waterloo, we would know very little of what really went on there.

Eyewitness testimonies are personal experiences much like Ensign Leeke's. However, there are often striking differences and inconsistencies in eyewitness accounts of the same event. Jurors listening to various eyewitnesses to a traffic accident, for example, may sometimes wonder how people seeing the same incident can report it so differently. This is particularly true if there has been a long time gap between the accident and the testimony in court. The jurors would most probably be convinced by the most confident witness, and they would, quite possibly, be mistaken. As many studies of eyewitness testimony have shown, confidence does not necessarily relate to accuracy.[2] Confidence comes from repeated retellings of a story, which may or may not be correct in the first place. You can be mistaken and confident as often as you can be correct and confident.

As time goes on, memory structures events, making them seem more "logical" and slanting them to put the narrator in a favorable light. People will often add explanations and commentaries to straightforward accounts to explain various things to their listeners.[3]

Because studies (which are discussed in more detail in the Appendix) have shown that the same basic processes are at work in the memories of college students, nonliterate Africans, and countless other groups of people around the world, we know that these same processes must have been at work in the minds of peoples in the past, in both individuals and groups, as they told their stories, formed their oral traditions, and carried these traditions down through time. What are the most important processes?

First, there is forgetting. Most forgetting is done shortly after an event, but the initial high rate of forgetting levels off, and the memory of an event stabilizes after a certain time.[4] In forgetting, memories become more general, details are lost, but more often the less important details. Stories get reduced to anecdotes. Numbers and names fare poorly.[5] But forgetting is selective. Details that define and validate an individual or group tend to get handed down. Earlier and less frequent events are remembered, but often telescoped toward the present; recent events are remembered. Those in the middle are forgotten. This is the "floating gap" found by Belgian researcher Jan Vansina.

Then there is embellishment, enhancement, or exaggeration. Implicational errors are introduced, explanatory glosses, narrative links so the story "makes sense" to listeners. Some items are exaggerated at the expense of others to give the story a certain effect.

Finally, there is assimilation or structuring. This is where real changes enter stories: the introduction of anachronisms, the fusion of similar incidents or people into one incident or person, or the transposition of details from one incident or person to another.[6] On fusion, "it sometimes seems as though memory tries to burden itself as little as possible. Instead of remembering separate items, it may be more economical to fuse them into a single general category."[7] In transposition retrieval errors come into play. Memory, especially as people age, is more and more "reconstructed" in one's mind, and retrieval errors—wrong event recalled or wrong time slice, that is, an error in the sequence of events—are the most common type of recall error.[8] Structuring will often occur to meet the contemporary needs of the community carrying down the story, which may change from generation to generation. If you can identify this type of structuring, you will learn something about how the group thinks about itself and what it wants to convey.

ORAL TRADITIONS THROUGH TIME

As messages evolve into oral tradition, they take different forms. Poetry is one of the most common. Epic poetry, like Homer's *Iliad* and *Odyssey*, is usually delivered orally for long stretches of time before it is ever written down. Alfred Lord and Milman Parry studied South Slavic narrative poetry and found, rather than simple memorization, that oral poets composed poems anew at each performance using stock scenes and descriptions and repeated phrases.[9] Some forms of oral tradition, such as tales and proverbs, are supposed to contain a good deal of improvisation at every telling. Others, often narratives, are supposed to be transmitted faithfully, as truthful accounts, even though the meaning of "truth" can vary from group to group.[10] This last category seems to fare best through time. In fact, the plot and the general sequence of episodes become set rather rapidly, and change after this is rather slight.

For example, looking at how an actual nineteenth century historical event was preserved in Hopi Indian oral tradition, one sees that a good deal of structuring (or assimilation) took place in the first two generations; particularly noticeable were changes in the relative importance of certain individuals as their actual influence within the tribe changed over time. One man, who had played only a minor role in the original telling of the event, grew more and more important in the story as it was retold, and as the man developed into a tribal leader. But another and much older

Hopi oral tradition, that of the coming of the Spaniards to Hopi territory in the early seventeenth century, had a good many elements that agreed with historical written records. From this one sees that most of the forgetting and structuring takes place in the first two generations after an event; beyond that, the process of change is very slow. This pattern mimics that of individual forgetting and retention over the short and long term, a not unexpected finding since traditions are "memories of memories."[11]

ORAL TRADITION AND ISRAELITE HISTORY

The longer narratives that eventually made their way into the first six books of the Hebrew scriptures were based on the foundational oral traditions of Israel, stories that for centuries defined them as a people separate and apart from their neighbors. Although there would be regional differences and variations in individual retellings, the essential contours of these foundational narratives would be reasonably stable through time. Biblical scholar Susan Niditch suggests that the Israelite oral traditions, passed down among the various tribes, took a fixed shape at the beginning of the monarchy and its centralized pan-Hebraic festivals in Jerusalem, much as the Greek epic poems took shape during pan-Hellenic festivals.[12] During such festivals oral retellings would become less and less variable and regional differences, such as the various northern and southern oral traditions, would have been to some degree flattened out. Many of the various strands of oral tradition came together during this period, *before they were written down*. The Levites may have been the principal tellers and keepers of these traditions, much as the Greek "rhapsodes" recounted the epics of Homer and Hesiod.[13]

Another biblical scholar, Frank Moore Cross, has suggested a largely poetic oral epic cycle that matured at the time of the Israelite league and was performed at cultic or pilgrimage festivals. Only later was it reformulated, passing through generations of editors, redactors, and copyists.[14]

HISTORICAL GOSSIP: HOW NATURAL EVENTS BECOME
MYTHIC TRADITION

As personal reminiscences pass into group tradition, they usually become mixed with the long-term manifestation of rumor known as "historical gossip." This sort of historical information may be extremely old. One account of a well near the Chad/Libya border into which the sun set each evening, heating the water in the well so that the people could cook their food, has been in existence for 2,500 years.[15] In North America, Klamath Indian oral tradition remembered, with remarkable geological accuracy, the eruption of Mount Mazama and the formation of Crater Lake, events that occurred more than 7,600 years ago.[16] A number of tribes in eastern New Guinea have oral traditions that remember two separate volcanic ash falls from two different eruptions of offshore volcanoes. Although many characteristics of these ash falls are remembered accurately, nearly all of the traditions have fused these two events, which occurred approximately 350 and 1,100 years ago, into a single "time of darkness."[17] Numerous other peoples throughout the world have oral traditions and myths that harken back to real natural phenomena such as earthquakes, tsunamis, and volcanic eruptions. These stories can shed light on both the historical context and the geological characteristics of such an event.[18]

THE NATURAL HISTORY OF THE EXODUS

Many of the stories found in the biblical books of Exodus, Numbers, and Joshua also contain these same natural phenomena: earthquakes, tsunamis, and volcanic eruptions. In particular there are three volcanic eruptions described in these ancient accounts. The first is the Minoan eruption of the Thera or Santorini volcano in the early seventeenth century before the Common Era or B.C.E., the second is a volcanic eruption in the northern Arabian volcanic shield at nearly the same time, and the third is another Aegean eruption nearly 180 years later. Over time the early people of Israel fused together and shifted these geological events in their oral traditions. Yet, once recovered, they serve as markers for the original time and settings of the stories. When these markers are combined with other geological, geomorphological, and paleoclimatological data, and with biblical scholarship, archaeology, and information from other ancient texts, many of the distortions and later alterations to the stories can be identified and set aside, and the original nature and sequence of the events which form the basis of the biblical accounts can be revealed.

This book will make the case that the Exodus narrative as we know it is the result of the oral transmission of these three separate volcanic events, the aftereffects of which were incorporated into Israelite oral history. Armed with an understanding of the ways in which oral history is constructed and transmitted, along with what the geological and archaeological record tells us about these volcanic events, we can plausibly reconstruct the actual events that underpin the Exodus narrative as we know it.

The Parting
of the Sea

CHAPTER ONE

Dating the Exodus

Actor Charlton Heston began his film career in 1950 on the steps of Chicago's Field Museum of Natural History playing Marc Antony in an adaptation of Shakespeare's *Julius Caesar*, the impressive pillars and white marble steps of the museum providing a highly effective stand-in for the Roman Senate.[1] Later he would go on to his most famous role, that of Moses in Cecil B. DeMille's epic film, *The Ten Commandments*. In this movie the biblical Exodus takes place during the reign of the pharaoh Ramesses II, of Egypt's Nineteenth Dynasty. In the year 2000, Field Museum Egyptologist Frank Yurco included this film in his class, "Exodus: The Egyptian Evidence."

EVIDENCE FOR THE EXODUS IN EGYPT

Frank Yurco (who died in 2003) was among a minority of Egyptologists who hold to the view that the Exodus actually occurred. Like many biblical scholars for the past several centuries, he cited what he believed was the most reliable part of the scriptural narrative: the names of the store-cities Pithom and Rameses in Exodus 1:11. This, Yurco asserted, pointed to the pharaoh Ramesses II, who reigned from 1279 to 1209 B.C.E.[2] Ramesses II's capital was at Pi-Ramesses, a close approximation of the biblical name. Pi-Ramesses was located in Egypt's eastern Delta region, thought to be the biblical "land of Goshen." Earlier pharaohs, those of the Eighteenth Dynasty, had their capital farther south, at Thebes or Amarna. Later pharaohs moved the capital to the city of Tanis. After this move the name Pi-Ramesses disappeared from common usage, as shown in the Bible where the name Tanis appears several times.

Yurco cited texts from the reign of Ramesses II to show that "'Apiru" (a term many scholars think relates to the biblical Hebrews) did indeed labor on the monuments of Pi-Ramesses. Most of the buildings of this and other Egyptian cities, he noted, were made of mud bricks such as those mentioned in Exodus 5. Unlike the earlier kings, Ramesses II did indeed build cities in the Nile Delta for storing his military supplies. The Pharaoh was also resident in his capital of Pi-Ramesses, and thus could have been physically accessible to Moses and Aaron, as the Bible account describes. Even the Red Sea crossing makes sense in terms of the city of Pi-Ramesses if the term Red Sea refers in fact to the Reed Sea (see chapters 4 and 10), since several marshy freshwater lakes filled with reeds were immediately to the east and northeast of that city. And, finally, Egyptian names in the Exodus account—Moses, Phineas, Hophni, Shiprah, and Puah—are "characteristic of the Ramesside era, less so in Dynasty XVIII and least of all in Dynasty XXVI."[3]

Other eminent scholars at a 1992 Brown University conference on the Egyptian evidence for the Exodus expressed their doubts about Yurco's position. Although archaeologist William Dever did agree that Egyptian historical evidence pointed to a thirteenth century B.C.E. date for the Exodus, he wondered how the newly escaped slaves could so quickly establish themselves in Canaan—for they appear as a distinct people, "Israel," on the famous Victory Stele of Merneptah of about 1207 B.C.E. Furthermore, the biblical account mentions the Israelites passing through the kingdoms of Ammon, Moab, and Edom. Ammon, Dever noted, was sparsely occupied in the thirteenth century B.C.E. while Edom and Moab were not yet established kingdoms.[4] Dever concluded that oral tradition may have preserved the memory of Canaanite groups in Egypt during the Hyksos period (seventeenth and sixteenth centuries B.C.E.) and their expulsion by the first pharaoh of the Eighteenth Dynasty, Ahmose, but that the true settling of Canaan by the early Israelites had nothing to do with the biblical Exodus or with the supposed wanderings in the wilderness and the subsequent conquest under Joshua, none of which fit any of the archaeological evidence.

Noted Canadian Egyptologist Donald Redford was even more pessimistic. Thirty years before he had pointed out that the Biblical names Pithom (*pr-'Itm* in Egyptian) and Rameses or Raamses were known only in the Saite period, that is, during the seventh and sixth centuries B.C.E.[5] Other concrete aspects of the Sojourn in Egypt and Exodus stories were likewise recent. As for an Exodus in the time of the Nineteenth Dynasty, he noted the total lack of any Egyptian evidence for a large population of Asiatics (that is, people from southwest Asia) in Egypt living in large measure unto itself during the entire New Kingdom (Eighteenth to Twentieth Dynasties).[6] Redford thought that the stories of the Sojourn in Egypt and the Exodus had their origin in the Canaanite (not Israelite) folkloric memory of the occupation of Egypt by the Hyksos, a people originally from southwest Asia.[7]

Another apparent nail in the coffin of a thirteenth century B.C.E. Exodus was provided by James Weinstein, who reviewed the archeological evidence from early twelfth century B.C.E. Israelite settlements and found hardly any evidence of Egyptian contact. Such contact would be expected from a people fresh out of Egypt. The only question that *really* mattered, Weinstein wrote, "is whether any (nonbiblical) textual or archaeological materials indicate a major outflow of Asiatics from Egypt to Canaan at any point in the XIXth or even early XXth Dynasty. And so far the answer to that question is no."[8]

Abraham Malamat of the Hebrew University in Jerusalem did discover an account of Asiatics leaving Egypt at the beginning of the Twentieth Dynasty. This group, in the first or second decade of the twelfth century B.C.E., was driven out of Egypt by the pharaoh Sethnakht after having been bribed with silver and gold to assist a rival political faction.[9] More than any of the other scholars at the conference, Malamat viewed the Exodus as the compression of a chain of historical or "durative" events telescoped into one "punctual" event.[10]

Both Dever and Weinstein pointed out the lack of archeological evidence for a thirteenth or twelfth century B.C.E. conquest of Canaan by Joshua.[11] William A. Ward summed up the consensus of the conference, and the mainstream of scholarly opinion, by noting that the Exodus could not be separated from the conquest under Joshua, and that "if there was no conquest, there is no need of an Exodus."[12] The archeological evidence is indeed unequivocal. Although there is much archeological evidence for the destruction of a number of Canaanite cities at the end of the Middle Bronze Age (starting about 1550 B.C.E.), there is little or none for their destruction when the conquest of Joshua would have occurred, if the Exodus had taken place during the Nineteenth Dynasty.[13]

DATING THE EXODUS FROM BIBLICAL AND OTHER ANCIENT TEXTS

More than twenty-five years ago a British scholar, John Bimson, attempted to solve this problem. First, he used the statement in 1 Kings 6:1 that the beginning of Solomon's temple (about 965–967 B.C.E. by modern calculation) took place 480 years after the flight from Egypt as a rough approximation of the actual Exodus date. Then he tried to move the dates for the end of the Middle Bronze Age forward more than one hundred years.[14] New archeological finds, however, as well as radiocarbon dates for the destruction layer of the walled city of Jericho, have shown this approach to be "fatally flawed."[15]

Earlier writers took a different approach to estimate the date of the Exodus, summing up the chronological information in the book of Judges and working backward from the reigns of kings David and Solomon. Using this method, in 1925 J. W. Jack estimated 609 years between the Exodus and the building of the first Israelite temple.[16] The most recent approach to determine the date of the Exodus involved computers. Using computer software to correlate the priestly cycles (taken from the Talmud), the lunar and solar cycles, and the jubilee years, E. W. Faulstich arrived at a date of July 31, 588 B.C.E. for the destruction of the Solomonic temple. Using the same method, he arrived at a date of 1421 B.C.E. for the conquest of Jericho, and by adding forty years to this figure, a date of 1461 B.C.E. for the Exodus.[17]

A much earlier writer, a first century C.E. Jew named Flavius Josephus, offered two dates for the Exodus. To counter the anti-Semitic claims of a writer named Apion, Josephus wrote a work entitled *Against Apion*, in which he quoted the third century B.C.E. Egyptian historian Manetho about the Hyksos, an Asiatic people who invaded and conquered Egypt in the first half of the second millennium B.C.E. Josephus equated the Hyksos to the Israelites to prove his own people's antiquity and stated that the Exodus had occurred 612 years before King Solomon built the temple.[18] In another work, *Antiquities of the Jews*, Josephus again used the 612-year figure along with a 466-year figure for the length of the the temple's existence. But elsewhere in *Antiquities* Josephus stated that the temple was started on the second month, 592 years after the Exodus, and also that the temple was destroyed 470 years, six months, and ten days after it was built. Combining the 592 years with the 470 years he went on to write that the temple was destroyed 1,062 years, six months, and ten days after the Exodus (and further that the Flood occurred 1957 years, six months, and ten days before the temple's destruction, and 3,513 years, six months, and ten days from Adam to the destruction).[19]

These sets of numbers apparently were from an ancient year-counting source, now lost.[20] This ancient source had at some point acquired the beginning of February as the starting point for each new year. Combining Josephus' year count of 1,062 years, six months, and ten days with the accepted date for the destruction of the first temple, the seventh or tenth of Ab, 586 B.C.E., produces an Exodus date of 1648 B.C.E., in early February.[21] However, if Josephus had actually made a twenty-year error *in the wrong direction* when he wrote 612 instead of 592 years, then the resulting figure—572 years between the Exodus and the break in the year count—would produce an Exodus date of 1628 B.C.E. As we shall see in chapter 3, this date is arguably the year of the Minoan eruption of Santorini/Thera. Josephus's time of year agrees with Egyptian harvest times as well (see chapter 4). The break designated as the start of the building of the temple is nearly a century too early for this event but would accord nicely with the destruction of the principal Israelite cult center at Shiloh, known to have occurred in the mid-eleventh century B.C.E.[22]

ORAL HISTORY, NATURAL EVENTS, AND THE STORY OF THE EXODUS

The modern-day oral historian would approach the Exodus story far differently than the literary scholar. First, the oral historian would give little weight to the fact that many people in the story don't have proper names, including Pharaoh—proper names often fall by the wayside in oral transmission. In the same vein, the names Pithom and Rameses, so important to literary scholars, would be treated with caution as possible later additions—anachronisms, a common feature of oral traditions. Second, the oral historian would give little weight to the number of years mentioned in 1 Kings 6:1, since numbers are likewise subject to great distortion. Moreover, this particular number is a multiple of forty and twelve, two ritual numbers for the early Israelites. An oral historian might pay a little more attention to the diverse numbers of years given for the rule of the judges, but some of them are recognizably ritual numbers as well. There is also the possibility of overlap for various judges in different parts or tribes of Israel, or missing periods, or other uncountable stretches of time.

Oral historians have often tried to use natural events to date traditional stories. But they have discovered that such events do not always stay attached to their original time and place.[23] A way to detect this problem is to look at the story as a whole. If an oral tradition does contain an extraordinary natural event (or a series of them), how intrinsic is the event to the story? Could the extraordinary event be moved or removed without changing the basic structure of the story? To put it another way, is it likely, in the context of the story, that the extraordinary event was added or moved?

The story of the Exodus contains a whole series of extraordinary natural or supernatural events. There is the burning bush, the ten plagues, and the parting of the waters. Certainly the plagues and some sort of miraculous event involving the drowning of the Egyptians are intrinsic to the story—without them there is no story, nor any reason to have such a story in the first place. It is worth noting that, in the ancient world, both the normal and abnormal occurrences of nature were held to be the works of the gods and goddesses. If something unusual had indeed happened, the people of the time, both Egyptian and Israelite, would have credited it to the working of divine authority.

NATURAL PHENOMENA AS EXPLANATIONS FOR THE EXODUS

With this in mind, in 1957 one ecologically minded scholar, Greta Hort, saw the plagues as disturbances in the ecology of the Nile, triggered by exceptionally strong July and August Nile flooding that brought down blood-red flagellates from the mountain lakes of Ethiopia, along with larger than normal quantities of the reddish sediments from the Abyssinian Plateau.[24] These flagellates, *Euglena sanguinea*, took oxygen from the river water, which killed the fish and brought on flies. This drove the frogs from the river not long before the high flood levels produced a lot of mosquitoes. Unfortunately, the frogs had contracted anthrax and spread it to animals and people, producing more of the plagues. Hail, coming in early February just before the barley harvest in the Egyptian Delta, destroyed the flax and barley, locusts blew in from Arabia, and a dust storm produced the exceptional darkness of the ninth plague.[25] Hort didn't explain the pillar of cloud and fire, however. In fact, large amounts of sediment from Ethiopia show up during low Nile floods, not high ones.[26] More importantly, the vicissitudes of the Nile floods and their effects would have occurred in other years and would thus have been regarded as ordinary events, whereas the Exodus portrays the water turning to blood as an extraordinary, one-time-only event. Moreover, how did such reasonably ordinary events get so closely connected in the minds of people (they supposedly happened over the course of most of a year) or come to be considered so extraordinary that they were remembered for centuries?

In a similar vein, archaeologists J. B. E. Garstang and his son John had earlier (in 1940) come up with the idea that the plagues were manifestations of a volcanic eruption that took place in the Rift Valley of central Africa. The Garstangs theorized that the central African lakes that are the sources of the White Nile were poisoned by Rift volcanoes, and the Nile brought the toxins north to Egypt, killing the fish and causing the earlier plagues. Another volcano, Mount Horeb, erupted in the land of Midian east of the Red Sea, and prevailing winds blew dust, steam, and ash over to Egypt, causing the hail and darkness plagues. An earthquake related to all this volcanic activity caused the sea to part and later to return and drown the Egyptians.[27]

Modern geological knowledge dispenses with this scenario, however. The volcanoes in central Africa are still active today, their effusive eruptions sending lava south into Lake Kivu, not northward to Lake Edward, which connects to Lake Albert, the source of the White Nile. The greatest danger humans and animals face from these basaltic shield volcanoes is through direct contact with the molten lava, or through asphyxiation from inhaling local pockets of carbon dioxide gas that form close to the ground. Only in the immediate vicinity of where the lava flows into Lake Kivu are fish parboiled, a bonanza to local fishermen.[28] Across the Red Sea, the effects of the volcanoes of Midian would only be felt locally, not as far away as Egypt.[29]

In 1964 a better candidate for the volcanic origin of the Exodus plagues emerged when A. G. Galanopoulos suggested that the Minoan eruption of the Santorini (Thera) volcano in the Aegean Sea was responsible for the plagues of the Exodus and the destruction of the Egyptian army in the Sirbonis lagoon on the northeastern coast of Egypt.[30] Despite being roundly criticized (but not usually by geologists and volcanologists), this idea became quite popular, although in fact archaeological remains indicate that the land spit over which the Israelites were said to have passed did not exist before the mid-first millennium B.C.E., well after any possible Exodus.[31] The connection between the Exodus and the Santorini eruption was discussed in Dorothy Vitaliano's 1973 *Legends of the Earth: Their Geologic Origins*, in Ian Wilson's 1985 book, *Exodus: The True Story Behind the Biblical Account*, and most recently in Elizabeth and Paul Barber's *When They Severed Earth from Sky: How the Human Mind Shapes Myth*.[32] Barber and Barber point out that parts of the Exodus story are quite characteristic of an ash cloud (their Group D) account of an eruption. In his book Wilson put the Exodus in the reign of the female pharaoh Hatshepsut, in accord with the theory of renowned Egyptologist Hans Goedicke.

THE EXODUS AND THE ERUPTION OF THE THERA VOLCANO

Goedicke made headlines in 1981 when he announced that the Exodus had occurred in 1477 and that the pursuing Egyptians had been drowned by a tsunami caused by the eruption of the Thera volcano.[33] In support of his theory he offered a new translation of Hatshepsut's Speos Artemidos inscription: "I annulled the former privileges [that existed] since [the time] the Asiatics were in the region of Avaris of Lower Egypt! . . . And when I allowed the abominations of the gods [i.e., these immigrants] to depart, the earth swallowed their footsteps! This was the directive of the Primeval Father [literally the father of fathers, Nun the primeval water] who came one day unexpectedly."[34]

This is a difficult text, and two other translators, Alan Gardiner and Donald Redford, have different endings. Gardiner's is: "Such has been the guiding rule of the father of [my fathers] who came at his [appointed] times, even Re"[35] and Redford's: "that was (?) the instruction of the father of the father[s] who comes at his regular times, *viz.* Re."[36] Redford does mention that the term "father of the father[s]" could mean a god, but an even more contentious item is whether the god, or the primeval water, came expectedly or unexpectedly. An unexpected appearance could refer to a tsunami, but an expected one certainly couldn't.

In 1992 Goedicke published a paper on the Thera/Santorini eruption which was in part a reaction to the scientific date for the Minoan eruption suggested at the Third International Congress on Thera and the Aegean World.[37] Like many other Egyptologists, he rejected this scientifically derived date of 1628 B.C.E. for the eruption and opted instead for a two-tiered Thera eruption, the first in the reign of Ahmose, first pharaoh of the Eighteenth Dynasty, and the second during the reign of Hatshepsut. Although there is no geological evidence for a two-tiered Theran eruption, Goedicke cited a mid to late first millennium B.C.E. *naos* from Saft el-Henna as support for a volcanic disaster in Hatshepsut's time. The naos is an inscribed rectangular block of granite, pointed at the top, with a large niche carved out of its front that once held the figurine of a god. Goedicke believes that the inscription on the naos is a mythologized history of the Eighteenth Dynasty from the time of Tuthmosis I to the beginning of Tuthmosis III's sole rule. This text describes an intense darkness that lasted for nine days; during this time the sea intruded inland.[38]

THE EXODUS FROM EGYPT AND THE CONQUEST IN JOSHUA

If Goedicke's reconstructions and attestations are correct, this event certainly has a good many similarities to the biblical Exodus. But there are also significant differences. The Eighteenth Dynasty pharaohs, and certainly Hatshepsut, lived much farther south in Egypt, in Thebes, not in the Delta. Moses and Aaron couldn't shuttle back and forth between Pharaoh and the Israelites living in the land of Goshen (undeniably located in the Delta) as they negotiated for the release of their people. Also, this Pharaoh had no sons, firstborn or otherwise, to die during the Passover; nor did she lead a pursuing army and drown in the sea of reeds. And lastly, and most tellingly, had the Exodus occurred in Hatshepsut's reign, it was not and could not have been followed forty years later by the conquest described in the book of Joshua.

In a very real way, the Exodus is connected to this conquest—as William Ward concluded at the 1992 conference, "if there was no conquest, there is no need of an Exodus." There are now radiocarbon dates on charred seeds from the only destruction level at Jericho that plausibly could have been associated with the Israelite destruction under Joshua. The average of these dates is 3311 ± 13 radiocarbon years BP (Before Present). Wiggle-matched to either the 1993 or 1998 tree ring calibration curve, this date falls in the middle sixteenth century B.C.E.[39] This is well before Hatshepsut's reign, before or at the very start of the Eighteenth Dynasty. If an Exodus from Egypt took place earlier, it would have occurred when the Nile Delta region was dominated by the people mentioned by Manetho, a Semitic-speaking people originally from southwest Asia, known to history as the Hyksos.

CHAPTER TWO

The Coming of the Hyksos

ARCHAEOLOGY AND THE HYKSOS

Genuine archaeology in Egypt goes back nearly 150 years, when serious excavators began to probe the dust-dry tombs along the middle and lower (or southern) reaches of the Nile. Relatively little archaeological work was done in the Egyptian Delta, however, for Delta sediments are nothing but mud, washed in by the yearly flooding of the Nile. Buried remains of ancient buildings and walls were made of mud brick, which could be distinguished from the encompassing mud only by feel—mud bricks were slightly more compact and sometimes had a slightly different color than the surrounding soil. Nonetheless, in the early 1970s, archaeologist Manfred Bietak of the University of Vienna, using modern excavation techniques, began digging in the northeastern part of the Nile Delta at the site of Tell el-Dab'a.

Bietak's team found excavation very difficult, for walls and buildings from higher levels in the soil cut into foundations from lower levels, and pits from more recent levels cut into older ones. The site was so large that the various excavated areas could not be correlated directly with one another, but only indirectly by comparing pottery types and subtle changes in building materials.[1] The Austrian team persevered, however, and what they found was worth the effort—Avaris, the ancient capital of the Hyksos.

The third century B.C.E. Egyptian historian Manetho described the Hyksos in the following way: "unexpectedly, from the regions of the east, invaders of obscure race marched in confidence of victory against our land. By main force they easily seized it without striking a blow; and having overpowered the rulers of the land, they then burned our cities ruthlessly, razed to the ground the temples of the gods, and treated all the natives with a cruel hostility, massacring some and leading into slavery the wives and children of others."[2] The first Hyksos king, Manetho reported, was named Salitis, who established his capital in Memphis. He also fortified a city in the Delta on the Pelusiac branch of the Nile to secure his territory against any invasion from the east. This fortified city, named Avaris, became the capital of later Hyksos kings. These Hyksos, as noted in the previous chapter, have sometimes been thought to have been, or at least included, the biblical Israelites living in Egypt just before the Exodus.

THE DELTA SITE OF TELL EL-DAB'A

Tell el-Dab'a (figure 2.1) consists of a mound or tell approximately five hundred meters in diameter, the remains of a town that once extended from the tell westward at least one kilometer to what was then the eastern bank of the Pelusiac Branch of the Nile. A freshwater lake that once formed the town's northern limit connected to the river by a feeder channel. Because the Pelusiac Branch of the Nile flowed northeastward to the sea, this feeder channel transformed the lake into an ideal inland harbor.[3]

The finds from Tell el-Dab'a have enormously expanded our knowledge of the era in Egyptian history known as the Second Intermediate Period. They have revealed that peoples from southwestern Asia (including the area that later became biblical Israel) settled in Avaris over 150 years before the advent of the Hyksos. Late in Egypt's Twelfth Dynasty (Stratum H—see table 2.1), sometime in the nineteenth century B.C.E., newcomers arrived at Tell el-Dab'a from Syria-Canaan, bringing with them a distinctive house form, Canaanite pottery of the Middle Bronze Age (MB) IIA type, and a large number of bronze weapons, notably the duck-billed ax seen in several Egyptian wall paintings.[4] While most of these newcomers were evidently urbanites (most likely from Byblos, along a part of the eastern Mediterranean known as the Levant), those living at Tell A, the "eastern suburb" of Avaris, were probably nomads or pastoralists. Brick enclosures there may have housed animals. The large number of weapons found in the male graves suggested that these newcomers were soldiers, recruited with their families by Egyptian rulers to guard the frontier.[5]

These pastoralists may also have come to the Delta because they were having trouble feeding their animals. Analysis of sediments in the Nahal Lachish indicates that central Canaan was experiencing erosion at this time, probably an indicator of dry climatic conditions.[6] In Egypt, the latter

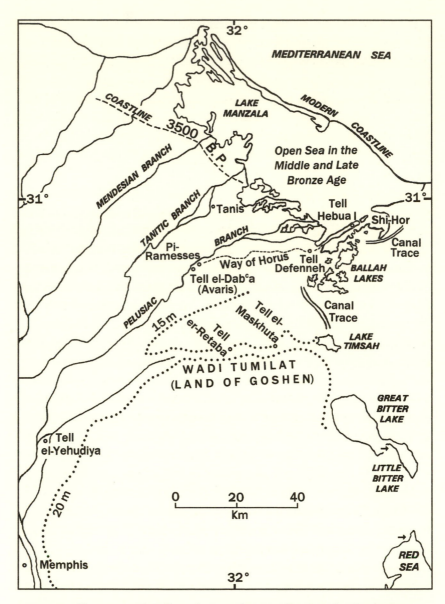

FIGURE 2.1. The eastern Nile Delta in the Second Intermediate Period and early New Kingdom, reconstructed from: Shafei; Bietak (*Tell el-Dab'a II*, *Avaris*); Sneh, Weissbrod, and Perath; Said; and Holladay. The 3500 (uncalibrated) B.P. coastline is from Coutellier and Stanley. The outlets for the Pelusiac Branch varied over time. The northern outlet shown here represents the one for 3500 B.P. (Coutellier and Stanley, figure 7c). The one further south is shown on Bietak (*Avaris*, figure 1) and Sneh, Weissbrod, and Perath (figure 1). Open circles on the present map denote sites. Arrows indicate the two land ridges normally under shallow water: between the Great and Little Bitter Lake, and on the northernmost extension of the Red Sea (Gulf of Suez). The area around Tell el-Maskhuta is thought to have been the biblical Succoth and possibly also Pithom (see text).

TABLE 2.1.
Strata at Avaris/Tell el-Dab'a, Tell A (from Earliest to Latest)

H	Egyptian late Twelfth Dynasty Middle Bronze Age (MB) IIA pottery
G/1–4	Egyptian early Thirteenth (or late Twelfth) Dynasty; terminated by a plague
F	Mid (or early) Thirteenth Dynasty; Canaanite temple and MB IIA/IIB pottery
E/3	First Asiatic rulers; start of Fourteenth Dynasty last evidence of Egyptian Thirteenth Dynasty; two large Canaanite temples; MB IIB or IIB1 pottery
E/2	Horses appear in this level; MB IIB or IIB2 pottery
E/1	Early Fifteenth Dynasty Hyksos rulers; large Canaanite population; MB IIB/IIC or IIB3 pottery
D/3	Fifteenth Dynasty Hyksos rulers; continuing population increase; MB IIC or IIB3 pottery
D/2	Late Fifteenth Dynasty Hyksos rulers; ends with destruction of settlement; MB IIC pottery

Note: These strata are found in one area of the excavation (Tell A) and are the ones most often cited in the literature. The central area of the site has a different set of stratigraphic designations. Manfred Bietak's dynastic labels for strata G/1–4 and F are listed first, followed in parentheses by William Dever's. Dever considered the Twelfth–Thirteenth Dynasty transition to be temporally equivalent to that of the MB IIA/IIB shift in Canaan.

part of the Twelfth Dynasty was marked by extremely high Nile flood levels due to the northward migration of the <u>Intertropical Convergence Zone</u> (ITCZ), which pushed the heavy monsoon rains north into the catchment areas of the Blue Nile and its tributary, the Atbara River, principal sources of the Nile. The northward migration of the monsoon regime in turn weakened the Mediterranean westerly winds, which carry most of the rainfall to Canaan.[7] Such conditions—drought in Canaan and high Niles in Egypt—could easily have compelled one or more groups of nomads in Canaan to seek more reliable pasturage in the Egyptian Delta.

The Twelfth Dynasty occupation at Tell el-Dab'a/Avaris was followed by a Thirteenth Dynasty palace in the central area. Nearby were tombs of high-ranking officials, treasurers or chief stewards. These officials were probably Asians, since the burials were accompanied by donkey sacrifices or the bones of sheep or goats, a characteristically Asian mode of burial. Such officials may have directed the trading caravans sent by land eastward to Asia and the seaborne trade between Egypt and the ports of northern Syria—Avaris was in fact a port. An Egyptian-made cylinder seal found in one of these tombs was inscribed with the image of the Syrian weather god Hadad/Baal Zephon, who would have been an important deity to sailors crossing the Mediterranean between Egypt and the Levantine ports.[8]

After this level the palace was abandoned, and the area was built over with small houses, most likely those of craftsmen. Craftsmen also lived in the eastern suburb area at Tell A, where small huts were built and round storage silos were erected in the enclosures. These levels (G/1–4, see table 2.1) were terminated abruptly by what appears to have been a plague, since there are mass graves at this level that apparently indicate a number of sudden deaths.[9] As a trading center, Avaris would have been exposed to peoples from many places, and diseases, such as bubonic plague and typhus, could have been brought in by ships or by donkey caravanners.

After the plague had passed through Avaris, larger houses with attached servants' quarters appear in the central section of the city—a clear sign of social stratification. The population at this time (stratum F) seems purely Canaanite. Tell A was deserted for a brief time for unknown reasons[10]; then the old huts were leveled and a sacred precinct built over them. Within this precinct a Near Eastern temple was erected. It continued to be in use, with some modifications, through succeeding strata.[11] Two limestone door posts, possibly from this temple, contain the name of King Nehesy (the name means "the Southerner" or "the Nubian"), who is listed on an ancient Egyptian king list (known as the Turin papyrus) as the first or second Fourteenth Dynasty king.[12] Other inscriptions suggest that Nehesy was devoted to the worship of the Egyptian god Seth, who like Baal Zephon had jurisdiction over storms. Some of the graves from this period contain the bodies of young girls buried at the feet of prominent men, a practice also found in Nubia's Kerma culture. Nehesy may have been the son of a Nubian princess, Tati, married to the first Fourteenth Dynasty king.[13]

In the succeeding stratum, E/3, a scarab of the Thirteenth Dynasty king Khaneferre Sobekhotep IV (ca. 1732–1720 B.C.E.; see table 2.2) was found in a grave. Scarabs, amulets shaped in the form of the dung beetle, are often inscribed with names in hieroglyphics on the bottom. Because they are quite common in ancient Egyptian sites and sometimes carry the name of a king, archaeologists have found them quite useful for dating. This Khaneferre Sobekhotep IV scarab is the last direct evidence of the Egyptian Thirteenth Dynasty at Tell el-Dab'a/Avaris and suggests, along with the Nehesy inscriptions, that the Asiatic Fourteenth Dynasty began about this time (see table 2.1).[14]

TABLE 2.2.
Partial List of Thirteenth Dynasty Rulers

Number	Nomen	Prenomen	Estimated dates (B.C.E.)
1	Sobekhotep I	Sekhemrekhutawy	1796–1793
2	Sonbef	Sekhemkare	1793–1789
3		Nerikare	1789
4	Amenemhet V	Sekhemkare	1789–1786
7	Amenemhet VI	Sankhibre	1783–1780
12	Sobekhotep II	Khaankhre	1775–1772
22	Sobekhotep III	Sekhemresewadjtawy	1749–1742
23	Neferhotep I	Khasekhemre	1742–1731
24	Sihathor	Menwadjre	1733
25	Sobekhotep IV	Khaneferre	1732–1720
26	Sobekhotep V	Merhotepre	1720–1717
27	Sobekhotep VI	Khahotepre	1717–1712
28	Ibiaw	Wahibre	1712–1701
29	Aya	Merneferre	1701–1677
30	Ini	Merhotepre	1677–1675
31	Sewadjtew	Sankhenre	1675–1672
32	Ined	Mersekhemre	1672–1669
33	Hori II	Sewadjkare	1669–1664
34	Sobekhotep VII	Merkawre	1664–1662
47		[...] mosre	
53		Se[...]enre	

Note: Numbers, names, and dates are from K. Ryholt, *Egypt*, tables 17 (p. 73) and 94 (p. 408) as amended by J. Allen, "The Turin Kinglist," in D. Ben-Tor, S. Allen, and J. Allen, *Seals and Kings* (pp. 50–51), who recommended removal of four names from Ryholt's list. The main source of this list is the Turin papyrus, which is badly damaged after Sobekhotep VII.

Horses make their appearance in stratum E/2. The more recent part of stratum E/2 and all of the succeeding stratum E/1 show evidence of a much larger population than before; according to Manfred Bietak, this reflects the creation of the Hyksos kingdom. American archaeologist William Dever, however, would place the advent of the Fifteenth Hyksos Dynasty in the succeeding stratum, E/1, while archaeologist Sturt Manning would place it even later, beginning in the subsequent stratum D/3.[15] The Hyksos proper comprise the Fifteenth Dynasty, six rulers said in the Turin papyrus to have ruled for 108 years.[16] The last two Hyksos kings, Apophis and Khamudy, are generally agreed upon, as is another ruler in the middle of the dynasty, Khayan. Several Asian rulers from the Second Intermediate Period are known only by their scarabs or seals: Yakbim, Ya'ammu,

Qareh, 'Ammu, Ma'aibre Sheshi, and Merwoserre Y'akub-Hr. These seals have been found in southern Canaan, in the Delta, along the Nile, and in Nubia.[17] Sheshi is the best candidate for Manetho's Salitis and Y'akub-Hr has been suggested as the second Fifteenth Dynasty king. In fact "Y'akub" (or Jacob) was a common western Semitic name in the Middle Bronze Age. The others are probably lesser kings under the early Hyksos rulers.[18] A scarab of Y'akub-Hr found in a tomb in northern Canaan was originally dated to the mid or late eighteenth century B.C.E., but has been redated and may coincide with the early Fifteenth Dynasty.[19]

The Hyksos Fifteenth Dynasty gained control over all of Egypt, if only for a short time, and seems to have installed a line of vassal rulers in Thebes, known as the Seventeenth Dynasty. Hyksos rule in Egypt was ended when Seventeenth Dynasty rulers mounted a series of military campaigns that resulted in the capture of Avaris and expulsion of the remaining Hyksos across the Sinai Peninsula back to Canaan.[20] It is this expulsion that has been claimed to echo through the folk-memory of the ancient Canaanites and to have found its way into early Israelite myth as the Exodus.

ASIATIC SITES OUTSIDE AVARIS

Although it was by far the largest, Avaris was not the only settlement of Asiatics in the Nile Delta during the Second Intermediate Period. About eighty kilometers southwest of Tell el-Dab'a (and twenty miles north of Cairo), Tell el-Yehudiyah (see figure 2.1) was excavated a century ago by British archaeologist Sir Flinders Petrie. Finds included burials, pottery, and scarabs, all from an obviously southwestern Asian population that lived there during the Second Intermediate Period. Petrie also excavated at the site of Tell er-Retabah in the Wadi Tumilat, south of the Pelusiac Branch of the Nile.[21]

THE WADI TUMILAT

The Wadi Tumilat is the remnant of an extinct channel that the Nile cut through the desert plateau and filled with sands and gravels in the last Ice Age.[22] Today the wadi extends from the present river course eastward to Lake Timsah and the Bitter Lakes (see figure 2.1). Until recent times the wadi, especially the low-lying ground in the western part, functioned as an overflow basin for the annual Nile floods.[23] When the Nile floods were high in ancient times the wadi would have been lush and fertile with a rich aquatic and faunal population. Seasonal and permanent water holes would have dotted the wadi's western and central parts, where limited cultivation was possible, although as a whole the wadi was suited more to the pasturing of flocks than to extensive agriculture.[24]

In its central part the Wadi Tumilat narrows at two points. At each of these points, along the northern flank of the wadi, is a tell, the western one known as Tell er-Retabah and the eastern one as Tell el-Maskhuta. The wadi as a whole has been extensively surveyed by a team headed by John S. Holladay, Jr. of the University of Toronto, and the largest site, Tell el-Maskhuta, has been excavated by them. Tell er-Retabah has also been the object of recent archeological excavation, but none of the results have been published.

Since the nineteenth century at least, the Wadi Tumilat has been equated with the biblical land of Goshen, the area occupied by the tribes of Israel during their sojourn in Egypt.[25] Its two principal sites, Tell er-Retabah and Tell el-Maskhuta, have been variously labeled as the biblical Pithom, Raamses, and/or Succoth. Pithom is from the Egyptian form *pr-'Itm*, "the house of Atum," while Succoth is derived from the Egyptian *T(k)w*.[26] But *pr-'Itm* is a Late Egyptian term; in its earlier usage it seems to have denoted not a town but open country, probably the open country in the vicinity of Tell el-Maskhuta. The term *T(k)w* also denotes a district rather than a town; like the "estate of Atum," it seems also to have been centered on Tell el-Maskhuta. In fact, the names Pithom and Succoth may both apply to the same area around and including Tell el-Maskhuta, one an older name (Succoth), and the other (Pithom) a more recent one.[27]

TELL EL-MASKHUTA

Tell el-Maskhuta is the largest archaeological site in the Wadi Tumilat, covering an area of 960,000 square meters, of which about two hectares produced traces of human occupation. Tell er-Retabah is less than half that size, 405,000 square meters.[28] Of the seventy other sites the University of Toronto team surveyed under the direction of Carol A. Redmount, twenty-one yielded remains from the Middle Bronze Age/Second Intermediate Period. Fifteen of these were no more than scatters of broken pottery sherds, and nearly all were found at the edges of ancient lakes or water holes where nomads and their herds would have camped for a short time. Five other sites in the wadi were tells; one was a burial ground. Two-thirds of these sites were in the central section of the wadi.[29]

Excavations at Tell el-Maskhuta revealed six phases of building and twenty-one burials.[30] The settlement began as a scattered, sparse, and mostly insubstantial occupation, founded on virgin fluvial deposits. Even in the earliest phases there were tombs, meandering perimeter walls, and circular structures commonly thought to be silos, although no traces of grain were ever found in them. Throughout the site's occupation, the buildings and perimeter walls were made of sun-baked mudbrick, manufactured without chaff (straw).

The first substantial structures were found in the third phase. The site became more densely populated as time passed and courtyards filled up with houses, but there were few luxury goods in any of the living levels. The inhabitants appear to have been quite poor. Activities at the site included spinning and weaving, pottery making, and some bronze-working.[31] Animal remains from the site included the bones of most domestic animals, principally sheep and goats (at least 70% of the animal remains recovered from all levels), cattle, donkeys, and pigs. The percentage of pigs increased through time, a clear sign that the people at Tell el-Maskhuta were staying longer in one place, since pigs don't travel well.[32] The plant remains, however, clearly showed that farming was done only in the winter months, when cereal crops—emmer wheat and barley—were grown. In the summer months there was no cultivation whatsoever.[33] In these months most of the occupants of the site must have moved into the open areas of the wadi, grazing their flocks by the lakes, wells, and water holes.

Most of the burials at Tell el-Maskhuta were in vaulted mudbrick tombs, although a few were in mudbrick-lined pits and a few more in simple holes in the virgin soil. Two young children were buried in imported Syrian jars, much like the child jar burials common at Tell el-Dab'a. Grave goods from the Maskhuta mudbrick tombs included some gold and silver jewelry, beads, amulets and scarabs, a few weapons, and pottery—mostly food bowls and beakers, suggesting food offerings to the dead. Later graves at the site were exclusively infants or subadults—apparently the adults were being buried away from the town.[34]

The assortment of pottery from the site, as well as from the other Middle Bronze Age sites in the central section of the Wadi Tumilat, is particularly interesting. One of the most popular ceramic forms in the early levels at Tell el-Maskhuta was the same sort of crude, handmade cooking pot found in the earliest level (stratum H) at Tell el-Dab'a. Those in the Wadi Tumilat, however, were much later in time.[35] As time went on at Tell el-Maskhuta, these handmade pots were replaced by wheel-made "hole-mouth" cooking pots.

Another common form was a wide, shallow "platter-bowl," usually thirty to fifty centimeters in diameter, that looks much like a modern wok. Fine and decorated pottery was rare to nonexistent. Cups usually had flat bases, quite different from the round-based drinking cups so common at Tell el-Dab'a.[36]

While most of the ceramics at the site reflect a generalized Syro-Canaanite heritage, many of the common Middle Bronze Age forms found in Syrian sites were not found at Tell el-Maskhuta, and the Wadi Tumilat's wide-mouth water jars were of Egyptian, not Syrian, origin. Carol Redmount, who studied the pottery extensively, concluded that the ceramic assemblage at Tell el-Maskhuta reflected an at least second-generation population of immigrant Asiatics whose Syro-Canaanite heritage had evolved through time and mixed with Egyptian traditions. She estimated that the Middle Bronze Age occupation at Tell el-Maskhuta lasted about fifty to one hundred years, probably closer to the latter figure.[37] She also suggested that the Middle Bronze Age sites in the Wadi Tumilat "should be seen as part of a political grouping," and that the people of Tell el-Maskhuta and the Wadi Tumilat were an ethnic subgroup within the greater Hyksos population.[38]

John Holladay has come up with a scenario to explain the occupation of the Wadi Tumilat within the framework of the greater Hyksos empire in Egypt.[39] The occupants of the Wadi Tumilat, he suggests, were settled there by a greater Hyksos authority to receive the winter-spring donkey caravans bringing incense and spices from South Arabia and the Far East, and gold and ivory from equatorial Africa. The water resources of the wadi would have been vital for these caravanners arriving from the desert, and Tell el-Maskhuta is strategically located to control access to the wadi from the east—from the Sinai and any trade routes headed south down both sides of the Red Sea. Maskhuta is about a day's journey from Tell er-Retabah. Caravans coming in from the Sinai Peninsula would stop at Maskhuta on one day and at Retabah the next, then proceed westward out of the wadi and turn north to Tell el-Dab'a/Avaris. From Avaris, goods would be shipped by boats to ports throughout the Eastern Mediterranean. This Wadi Tumilat trade route would be particularly important at times when the rulers of Upper (southern) Egypt shut off communication via the Nile River, so that travel overland from Arabia, the Horn of Africa, or via the oases of the Western Desert would have been the only way for goods from the south and southeast to reach the Delta.

Holladay cites specific features found at Tell el-Maskhuta to support his scenario: (1) the relatively rich grave goods (brought in by the caravanners and traded, perhaps, for meat and milk) in the burials, so different from the humble items found in the living areas of the site, and (2) the over-sized cooking facilities in the site's occupation areas and the multiple jar emplacements stretching across compounds—think of cafeterias set up to feed the caravanners. He also notes the real degree of military prepared-ness evident in the weapons found at the site, and the importance of don-keys, which often were sacrificed at the tombs of important males, both here and at Tell el-Dab'a.[40]

This scenario sees the people of Tell el-Maskhuta and the adjacent sites in the central section of the wadi as a small community of pastoralists (about three thousand people, if nineteenth century Bedouin populations in the wadi are any guide[41]) settled in the Wadi Tumilat by the Hyksos rul-ers, living in small towns and hamlets in the winter and early spring to grow wheat and barley and make the cloth, pottery, weapons, and other items they would need for the rest of the year, all the while guarding the frontier and servicing the donkey caravans coming in from the east. In the summer they would leave the towns and hamlets to pasture their flocks be-side the wadi's water holes, occasionally hunting wild game and waterfowl. This picture fits well with Redmount's ideas of the wadi's people as a dis-tinct sub-group within the greater Hyksos population; archaeologically, it would be most appropriate to the later levels (phases 5 and 6) at the site.

DATING OF TELL EL-MASKHUTA AND THE WADI TUMILAT SITES

Tell el-Maskhuta scarabs date from the later part of the Thirteenth Dynasty (Khaneferre Sobekhotep IV) through the first part of the Hyksos Fifteenth Dynasty, equivalent to the period known as Middle Bronze (MB) IIB.[42] James Weinstein, who studied the scarabs, says that they are similar time-wise to those found in tomb groups III–IV and IV–V at the site of Jericho in southern Canaan and that their equivalent levels at Tell el-Dab'a are E/1 and its succeeding level, D/3.[43] He first estimated that the scarabs from Tell el-Maskhuta dated from 1750 to 1625 B.C.E. More recently, based on both the pottery and the scarabs, he wrote that occupation at Tell el-Maskhuta ended slightly earlier than occupation at Tell el-Yehudiyah, which he thinks ended around 1575 B.C.E., while Carol Redmount estimated the occupa-tion of Tell el-Maskhuta from about 1700 to 1600 B.C.E.[44]

In terms of the pottery, John Holliday limits occupation at Tell el-Maskhuta to an even narrower range: "all of our pottery would fit com-fortably within the limits of (at the earliest) late Stratum E/1 and (probably at the latest) Stratum D/3 at Tell el-Dab'a. . . . Probably we lack the earliest

E/1 material and the latest D/3."[45] Manfred Bietak assigns dates of 1620 to 1560 B.C.E. to the E/1–D/3 strata at Tell el-Dab'a.[46] Bietak's estimates would mean that occupation at Tell el-Maskhuta lasted no longer than thirty to forty years, a timespan that makes no sense in terms of Weinstein's and Redmount's estimates.

The scarabs from Tell el-Maskhuta include Thirteenth Dynasty types that extend back to Tell el-Dab'a level E/3 (see table 2.1). Moreover, William Dever has identified some late MB IIA and transitional MB IIA/IIB pottery forms from Tell el-Maskhuta.[47] Therefore, the occupation at Tell el-Maskhuta arguably began as early as the E/3 stratum and lasted into stratum D/3, a timespan more in keeping with the dates and length of the occupation estimated both by Redmount and Weinstein.

This discordance highlights a raging debate between two opposing camps of archaeologists. Many archaeologists studying the Middle Bronze Age in Canaan have a whole series of dates that are far older than the dates Manfred Bietak assigns to the equivalent strata at Tell el-Dab'a. William Dever, for example, dates the E/1–D/3 layers at Tell el-Dab'a to 1675–1575 B.C.E.[48] All of these estimates are based on approximations of the length of each pottery or occupation phase. Unfortunately, the only radiocarbon dates published for any level at Avaris/Tell el-Dab'a have too wide a range (150 calendar years for one from late in level G, 113 radiocarbon years, the equivalent to 194 calendar years, for the average of two other dates) to resolve the dating controversy.[49]

In contrast, Sturt Manning, an archaeologist whose speciality is Bronze Age pottery from Cyprus, has closely studied correlations between wares imported from that island to Avaris, Tell el-'Ajjul in southern Canaan, and Tell el-Maskhuta. From his work he has concluded that the Fifteenth Dynasty is represented by strata D/3 and D/2 at Avaris and that the earlier part of the Hyksos rule is contemporaneous with Late Minoan IA, which radiocarbon dates reveal ended between 1620 and 1603 B.C.E. Consequently, Manning has stratum D/3 extending back from about 1600 B.C.E. to the mid-seventeenth century B.C.E.[50] This date range would be more in keeping with Redmount's and Weinstein's (first) dates for the occupation of Tell el-Maskhuta.

ABANDONMENT OF THE WADI TUMILAT SITES

There is a far greater problem with Tell el-Maskhuta than its date range, and that is explaining why the site, and *nearly all of the other Middle Bronze*

Age sites in the Wadi Tumilat, were abandoned before the end of the Hyksos rule in Egypt. Except for a few occasional "squatters," occupation at Tell el-Maskhuta ceased before the end of what is equivalent to the D/3 layer at Tell el-Dab'a; Hyksos rule in Egypt lasted through the end of the ensuing D/2 stratum there.[51] People did not resume living at Tell el-Maskhuta until the Saite Period (seventh century B.C.E.), a gap of about a thousand years. Only at Tell er-Retabah, closer to the western end of the Wadi Tumilat, did occupation occur in the New Kingdom's Nineteenth Dynasty.[52]

Holladay suggests that, as the Hyksos came to dominate all of Egypt, transport of luxury goods by boat up the Nile to Avaris supplanted the land route across the Wadi Tumilat.[53] Certainly carvings with the names of the Hyksos rulers Khayan and Apophis were found in Upper Egypt not far from Thebes, indicating Hyksos power in the south, but this explanation does not agree with the finding of numbers of seals of earlier Hyksos rules along the Nile and in Nubia, an indication of trans-Nile trade.[54] With or without the overland trade, the wadi should have remained a place for winter cereal farming and a prime pasturage for the flocks of sheep and goats that represented nomadic people's real wealth, since Nile flood levels were normal throughout this period.[55] Besides the possible cessation of overland trade, Holladay concedes that there is "no ready explanation" for the discontinuance of Hyksos occupation in the Wadi Tumilat.[56]

The question then needs to be asked: what happened in stratum D/3 that could have resulted in the abandonment of virtually all of the Wadi Tumilat sites? To put it another way, why did all the people living in the Wadi Tumilat leave it, sometime in stratum D/3, and never return?

Interestingly enough, by accepting the dates of Redmount, and Weinstein's original dates for Tell el-Maskhuta, and Manning's dates for the D/3 stratum at Avaris/Tell el Dab'a (from the mid-seventeenth century to about or slightly before 1600 B.C.E., which are correlated with radiocarbon-derived dates in the Aegean) the D/3 stratum is also the level in which the Minoan eruption of the Santorini volcano occurred.[57]

CHAPTER THREE

The Minoan Eruption

Over the last two decades or so a wealth of new scientific information has become available about the Minoan eruption of the Santorini/Thera volcano. Other scientific research sheds new light on the nature of large-scale eruptions in general and the effects these eruptions have on people, plants and animals, and the environment. With this information in hand, this chapter will describe the Minoan eruption and its probable effects, particularly the effects of its tsunami and airborne ash clouds. Then, the next chapter will go on to compare the eruption and its effects to the plagues described in the book of Exodus. Let us start by looking at when the eruption actually took place.

THE CONTROVERSY OVER THE DATING OF THE
MINOAN ERUPTION

The Minoan eruption of Santorini/Thera is a key marker for the Bronze Age archaeology of the eastern Mediterranean world. For most of the twentieth century, archaeologists placed it at about 1500 B.C.E. As radiocarbon dating techniques were applied to Bronze Age archaeological material, however, this date appeared to be over one hundred years too young.[1] When tree-ring chronologies and evidence from the Greenlandic ice cores appeared to support the radiocarbon dates, the dispute over the true date for the eruption grew heated and intense. Recent analysis of volcanic glass from a layer in the Greenland ice cores previously thought to mark the Santorini eruption showed that this layer actually marks an eruption of Aniakchak, an Alaskan volcano; the seventeenth century B.C.E. growth spurt evidenced in the Anatolian tree-ring chronology can also be linked to this Aniakchak eruption.[2] There thus remain two camps—many earth scientists, tree-ring experts, and some archaeologists who support the earlier, seventeenth century B.C.E. date for the eruption, and those who support, on historical grounds, a date of 1500 B.C.E. or even slightly later.[3]

The dates of Egypt's Second Intermediate Period and the beginning of the Eighteenth Dynasty are at the very heart of this debate, since the Egyptian chronology is the basis for many other archaeological chronologies in the eastern Mediterranean. Archaeologist Sturt Manning, mentioned in the previous chapter, argues that the seventeenth century B.C.E. eruption date for Santorini/Thera can fit with a higher Egyptian chronology and the "early" Aegean chronology. It cannot fit with the low chronology proposed by Manfred Bietak.[4]

The most direct geochronological evidence for the seventeenth century B.C.E. date for the Minoan eruption comes from radiocarbon dates on samples taken from the destruction layer on the island of Thera itself. The best of these, on fully carbonized seeds found in sealed jars buried by the eruption, have been radiocarbon dated at 3344.9 ± 7.5 ^{14}C years B.P. Calibrated to the most recent tree-ring curve, these results produce a range of 1660–1613 B.C.E. at a 95.4% confidence level, with the most likely subrange being 1639–1616 B.C.E. A second report dates an olive branch found in the volcanic deposits of Thera. Radiocarbon-dating and wiggle-matching the rings on this branch, which was by all accounts alive up to the time of the eruption, produce a calendar date range of 1627–1600 B.C.E. at the 95.4% confidence level. However, if there is only a 25% error in the counting of the rings on this branch, the radiocarbon date range is 1635–1591 at the 95.4% confidence level, and the overlap with the dates from the carbonized Theran seeds is 1635–1616 B.C.E.[5]

Three of the Greenlandic ice cores, the Dye 3 core in southern Greenland and the GRIP and GISP2 cores in northern central Greenland, record acid spikes at 1622 (Dye 3), 1618 (GRIP), and 1618 (GISP2, corrected date) B.C.E.[6] Given that these dates are plus or minus several years, they are functionally identical and probably signal the Minoan eruption. The eruption date indicated by these ice core acid spikes may be six to eight years too low, however, since the radiocarbon date on the Anatolian tree-ring growth spurt linked to the Aniakchak eruption is six to eight years higher than its acid spike date of 1645 B.C.E.[7] If this is the case, then the date of the Minoan eruption would fall squarely at the time when absolute tree-ring chronologies from North America and Europe record a growth anomaly thought to be the product of a volcanic eruption—1628 B.C.E.[8] As noted in chapter 1, this date is closely related to the date Jewish historian Josephus gives for the Exodus.

Santorini is a subduction-zone volcano—it owes its existence to the subduction of the African tectonic plate beneath the Eurasian plate. As the subducted slab goes deeper and deeper into the earth, it releases water, which by lowering the melting point of part of the earth's mantle known as the asthenosphere causes it to partly melt. Because this melted material is less dense than the rest of the asthenosphere, the melt (or magma) rises to relatively shallow levels inside the earth's crust, where it may stay for years or even centuries in one or more magma chambers. If something happens to increase the pressure inside the magma chamber so that it becomes greater than the pressure of the overlying rocks, a volcanic eruption, like the Minoan eruption, occurs.[9]

The Minoan eruption of Thera is the second largest explosive eruption in the past four millennia, a Volcanic Explosivity Index (V.E.I.) 7 or more cataclysm that ejected the equivalent of sixty cubic kilometers of dense rock into the ocean and the atmosphere to heights of thirty-six to thirty-eight kilometers, well into the stratosphere.[10] In modern times, only the 1815 eruption of Tambora in Indonesia was larger, and that eruption was so immense it produced "the year without a summer" throughout the Northern Hemisphere.

Several months before the Minoan eruption, Santorini experienced an earthquake strong enough to damage buildings. It was followed by a precursor ashfall (called the BO_0 phase by geologists) a few weeks or months before the eruption itself. This preliminary volcanic activity apparently caused the island's inhabitants to flee, since no evidence exists that people were killed on Thera as they were at Pompeii.[11]

The first phase of the eruption (BO_1) itself is termed a plinian phase or eruption, named after Pliny the Younger, who as a young man witnessed and described the eruption of Vesuvius in Italy in 79 C.E. In this phase, volcanic tephra and ash shot up into the air with increasing violence and intensity. Up to seven meters of rose-colored, iron-rich pumice and ash were deposited on the islands of the Santorini archipelago, in patterns that indicate a strong wind blowing toward the southeast at the time. Tephra from this phase was also found in sea cores southeast of Santorini. There was no interaction between the erupting magma and sea water—the magma was fragmented and discharged by its own exploding gases.[12] Pumice falling into the sea would have formed enormous rafts that floated around the eastern Mediterranean. When Krakatoa erupted in 1883, pumice rafts transported skeletons and trees around the Indian and Pacific Oceans.[13] At Santorini, this plinian phase lasted about eight hours but did not generate any tsunamis.[14]

In the second phase (BO_2), erupting vents opened in the sea south of the first vent and sea water interacted with the magma. This caused violent explosions that pulverized the magma and exploded large blocks of it onto Thera, along with about twelve more meters of ash and pumice. Pyroclastic flows and surges (also called *nuées ardentes*) occurred during this stage, which lasted at least an hour and perhaps as long as a day; similar surges killed thousands of people in Pompeii and Herculaneum when Vesuvius erupted in 79 c.e.[15] Tsunamis were likely generated as these massive surges entered the sea, but particularly toward the north, south, and southeast. Evidence from Palaikastro in northeast Crete shows that a massive tsunami or tsunamis from the Minoan eruption was directed to the southeast, directly toward Egypt.[16] Similar pyroclastic flows entering the sea during the eruption of Krakatoa produced tsunamis that killed thousands of people.

In the third phase (BO_3) there is again clear evidence of sea water mixing with the magma. About fifty-five meters of white pumice and ash, interbedded with larger rocky material, were deposited on Thera, possibly from a new vent to the west of the original one, as hot pyroclastic flows welled out of the caldera and down into the valleys. This phase lasted about a day, and huge tsunamis would have been generated wherever pyroclastic flows entered the sea. These tsunamis would have been channeled to the west and southwest by now-opened fault blocks.[17] As the pyroclastic flows spread out they combined with—and heated—the air above them, forming a buoyant column containing vast quantities of eruptive material. This is what is called a co-ignimbrite eruption column, and it was this column, and not the plinian eruption plume, that produced most of the Theran tephra that has been recovered in eastern Mediterranean sea cores and from deposits to the northeast of Santorini as far as the Black Sea.[18] Toward the end of this third phase the caldera, already subsiding, began its final collapse.

Most scientists believe that the Minoan eruption had a fourth and final phase (BO_4), in which ignimbrites and other sediments were deposited on the broad coastal plains of Thera and Therasia.[19] The final stages of caldera collapse created more tsunamis toward the west-southwest, forming massive deposits called homogenites on the seafloor of basins to the west and south.[20]

TSUNAMIS

Tsunamis are enormous waves that reach from the surface of the ocean to the deep seafloor and travel at speeds of up to eight hundred kilometers per hour in the open ocean. Once a tsunami reaches shallow waters off-shore, the bottom of the wave slows down and drags at the upper part, causing the wave to bunch up to a great height just before it crashes onto the land with an awesome destructive power, bringing debris from the ocean and the ocean floor with it.[21] Tsunami deposits from the Minoan eruption have been found on Thera itself, in northeastern Crete, and in two locations along the coast of western Turkey. In Israel, a Minoan tsunami may have caused the cliff collapse at Tel Michal, in its Middle Bronze IIB layer.[22]

ERUPTION CLOUDS AND TEPHRA DISPERSAL

Clouds from a volcanic eruption may be carried horizontally for thou-sands of kilometers, but satellite studies of modern eruptions show that 75–90% of erupted tephra falls to ground in the first thirty-six hours, mostly as fine ash.[23] The distance the cloud travels is dependent on both the amount of pyroclastic material ejected into the atmosphere and the wind velocity. Tephra clouds travel at markedly different speeds, from twenty-five to one hundred kilometers per hour, with higher-altitude clouds traveling faster.[24] These clouds may follow straight trajectories or curve around storm systems, and their direction or rate of drift may vary with position or altitude. Varying wind direction may produce a broad, perhaps lobate, mantle, and the more prolonged the eruption, the greater the possible variation.[25]

Minoan tephra deposits covered an immense area, estimated at 2–2.2 million square kilometers.[26] Distribution of these deposits indicates a broad rather than a concentrated pattern: in sea floor sediments from the eastern Mediterranean, on the Greek islands, in Anatolia, and even in de-posits from the Black Sea.[27] Theran glass shards have been found in sedi-ments cored from the Nile Delta,[28] and recent excavations in the area of Tell Hebua I (see figure 2.1) near the Mediterranean coast has uncovered houses, military structures, and tombs encased in ash, along with frag-ments of pumice.[29]

In the first, intense, plinian phase (BO_1), deposits on the Santorini ar-chipelago show that the wind, at least at lower altitudes, was strongly from the northwest and would have blown the tephra clouds southeast. The predominant summer winds do blow in this direction in the Aegean.[30] These winds alone would not explain the broad distribution of the Theran tephra, however, which includes deposits in Anatolia and the Black Sea, well to the northeast of Santorini. The Anatolian tephra deposits are better explained by west and southwest winds that are most common in the winter and spring months, not in the summer.[31] One suggested solution to this timing problem is that lower-altitude summer winds were responsible for the south- and southeast-directed deposits and that the northern and eastern tephra deposits were produced by fallout from the upper troposphere, where the jetstream runs in a generally easterly direction over the Aegean and north to northeast over Anatolia.[32]

Another possible solution involves the complex weather systems that move through the eastern Mediterranean in the winter months. From December through April, dry cold air at high altitudes, originating in the Arctic, moves south over Europe. Reaching the Mediterranean, these high-altitude air pockets come in contact with the warmer, wetter air over the sea and form depressions over the Gulf of Genoa, the north Adriatic, and the western Aegean north of Crete.[33] The Aegean depressions cause cyclonic storms that last four to seven days and move generally southeastward to Egypt or curve northeastward over Anatolia and into the Black Sea. Some ten to twelve storms strike the coast of northeast Egypt each winter.[34] If Santorini had erupted in the mid or late winter months, one such storm could have carried the airborne material from the earlier eruption stages southeast while another, following closely behind, curved northeast across Anatolia in the final stages of the eruption, carrying material from the co-ignimbrite eruption column with it. The latter storm could also have produced the torrential rain thought by some scientists to have helped form the deposits of the fourth (BO_4) phase (the summer winds are dry).[35] These weather conditions would have assured the dispersal of the Santorini tephra as far south as the Egyptian Delta and as far north as the Black Sea (see figure 3.1).

How much ashfall remains in place centuries after an eruption, even a very big one, is quite problematical. One of the best examples, because it combines contemporary accounts with modern measurements, is the February 19 to March 5, 1600 C.E. eruption of the volcano Huaynaputina in what is today southern Peru. This eruption is recorded in historical sources, the Northern Hemisphere tree-ring record, and the Antarctic ice cores.[36] The Huaynaputina eruption released an estimated 19.2 cubic

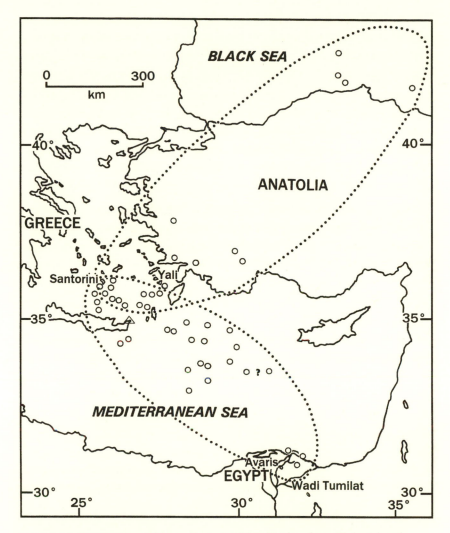

FIGURE 3.1. Distribution of airborne Santorini (Thera) tephra, from: Ninkovich and Heezen; Watkins et al.; McCoy ("Areal Distribution,"); Sullivan; Guichard et al.; and Eastwood et al. Open circles mark deep sea and other drill cores that produced Minoan eruption tephra. Triangle marks location of tsunami evidence in northeast Crete. The two ovals outline the pattern of the two wind or storm tracks.

kilometers of rock (compared to the sixty cubic kilometers estimated for the Minoan eruption). Eyewitness accounts record that tephra fell in Lima, about eight hundred kilometers away, and on a ship about one thousand kilometers out to sea. Historical accounts also report that at least one meter fell north and west of the city of Arequipa. Today, one-centimeter-thick layers of Huaynaputina's tephra can be found no farther than two hundred kilometers northwest of the eruption, not 800–1000 kilometers, and the deposits in the vicinity of Arequipa are only ten centimeters thick at best, not one meter thick.[37] These measurements indicate that tephra in most of the distal ashfall area has now disappeared and that there has been a 90% reduction in the thickness of the original tephra deposits in only four hundred years.

In another example, the Krakatoa eruption of 1883 (with a V.E.I. of 6.5) produced ash that fell on the island of Timor over thirteen hundred miles away, in the Indian Ocean as far as two thousand miles away, and even on ships off the Horn of Africa, 3,700 miles away. Numerous eyewitness accounts describe the extensive amounts of ash that fell on the nearby islands of Java and Sumatra and in the waters around these islands, especially in the Sunda Straits.[38] Yet ash from this eruption, "whose atmospheric extent and effects are so well known, shows up little if any in deep-sea cores."[39] In light of such modern analogs, it seems safe to say that the often-quoted figure of 50% reduction in present-day Santorini tephra layers from their original thickness[40] is a minimal estimate and that quite thin tephra layers, or even restricted finds of Theran glass, are remnants of much thicker deposits.

It is now known that ice can play a key role in eruption clouds, particularly in clouds from sea-level volcanoes such as Santorini.[41] As water vapor rises in the atmosphere it freezes, and the resulting ice will surround fine particles of volcanic ash to form icy ashballs. Icy ashballs carried in an eruption cloud will either fall as hail, or, if it is warm enough, will melt or evaporate before reaching the ground.[42] These ice/ash particles clump together and fall out at significant distances from the eruption. The 1992 eruptions of Spurr volcano (V.E.I. = 3) in Alaska produced secondary maxima ashfall areas 150–350 kilometers downwind of the volcano, while ancient ash deposits in the Great Plains of North America record secondary maxima an order of magnitude thicker than what would be expected from modern ashfall studies and, in at least one case, occur 1,400 kilometers away from the original eruption.[43]

Along with ice, sulfur dioxide is also a major component of many eruption clouds. The sulfur dioxide may be released selectively before much of the ash, and once in the air it forms sulfuric acid, while fluorine and chlorine, also commonly present, form hydrofluoric and hydrochloric acids, respectively. These acid aerosols are adsorbed onto the surfaces of the erupted fine ash particles or scavenged by ice. Given the high amounts of chlorine and especially sulfur in known deposits of Santorini tephra, and the estimated amount of sulfur released in the Minoan eruption,[44] the eruption clouds, particularly those formed in the first stage of the eruption when much of the sulfur was probably released, would have carried highly toxic concentrations of these acid aerosols. Adsorbed onto ice and ash surfaces, these aerosols would have fallen to earth with the ash and the ice/hail/rain. Since the smaller particles have more surface area in proportion to their volume, they will carry higher concentrations of these acids, and also be carried farthest.[45] As volcanic hazards expert Richard Blong states: "scavenging of eruption clouds by rainfall may occur at a variety of times and distances from the vent at distances of hundreds of kilometers. Similarly, the concentration of fluorine and other potentially toxic substances on smaller tephra grains raises the possibility of adverse hazards at distances quite remote from the erupting volcano."[46]

EFFECTS OF ERUPTION CLOUDS AND TEPHRA FALL

Most of these ashborne acids will cause blisters and burns if they come into contact with the skin and lips, particularly if the skin is wet (as in a rainstorm). They will also cause severe eye irritations. Fluorine poisoning will cause lesions in the nose and mouth and on the legs and cause hair to fall out. Breathing hydrofluoric acid will corrode mucous membranes in the lungs, and the tephra itself will coat these same mucous membranes when people or animals breathe it.[47] The eruption of Mount St. Helens in 1980 caused a two- to threefold increase in acute asthma and bronchitis cases in eastern Washington hospital emergency rooms in the days after the eruption and for the following month.[48] Cattle, horses, sheep, and goats that breathed in volcanic ash during the eruption of Paricutin in Mexico in 1945 died months later from the effects of ash-mucous coatings in their lungs.[49] Animals such as sheep that eat vegetation close to the ground are more affected than those that graze at higher levels. Reindeer, another close-grazing animal, have been affected by as little as twenty-five millimeters of tephra fall.[50]

Fish too will be affected by tephra fall, particularly if it is accompanied by acid rain. Freshwater fish are killed by as little as thirty millimeters of tephra fall if it is highly acid (a pH of less than 5). In the eruption of Mount Spurr in Alaska in 1953, three to six *millimeters* of tephra dropped the pH of the public water supply of Anchorage to 4.5 for a few hours and caused a great deal of turbidity. Hundreds of dead fish were found in areas that received only ten millimeters of tephra from the 1979 eruption of Karkar volcano in Papua New Guinea.[51]

Plants will be affected by the weight of the tephra and even more by acid burns, particularly if the acid comes with rainwater. Leaves will turn brown, wilt, and fall off. Volcanic dust coated with a film of hydrochloric acid burned plants around Mayon volcano in the Philippines in 1928.[52] Of the cereal grains, wheat and barley are the most sensitive to this kind of damage. Acid rains and gases killed vegetation 480 kilometers from Katmai-Novarupta volcano in Alaska when it erupted in 1912, and areas with less tephra fall were affected more severely than those with a larger ashfall.[53]

The effects of eruptions and tephra fall on people are no less profound. Many historical accounts testify to the enormous effects of eruptions and ashfalls, and modern psychological studies show widespread mental health problems after eruptions, little different from those that follow other natural disasters.[54] The first reaction to imminent disaster is usually denial, not believing that the disaster will happen. Before the eruption of Mount St. Helens in 1980, the news media gave much space to an eighty-three-year-old man named Harry Truman who refused to leave the danger zone (and was killed in the eruption).[55] Other people may experience anxiety that turns to terror when disaster strikes. At this point, action is usually taken to reduce losses. Often this action is religious. The veil of Saint Agata was used by the people of Catania to protect their city from eruptions of Mount Etna in 252, 1408, 1444, and 1669. When the Paricutin volcano in Mexico formed in 1943, the local villagers believed that a sacred image in the church of San Juan Panangaricutiro would save their village, to no avail.[56]

In groups with a tradition of human sacrifice, such as some in Meso and South America, infants, children, or young girls were sacrificed to volcanoes. At Huaynaputina, "the nicest young girls, the best animals, and the prettiest flowers" were sacrificed to the volcano.[57] Other groups practiced animal sacrifice. During the time of darkness caused by the mid-seventeenth century Long Island eruption, the tribes of Highland New Guinea sacrificed pigs: "The third day [of darkness] was like the first two and now the people decided they must do something to make it light again. They killed a white-skinned pig." Another group killed a black dog and a black pig to make it light again.[58]

One common urge during a natural disaster is to flee. More sedentary people would be less inclined to flee than those—such as hunters and gatherers or the nomadic pastoralists in the Wadi Tumiat—who are less sedentary.[59] Northern Athapascan hunters, living a marginal existence hunting and fishing in the cold northern forests of British Columbia and the Yukon Territory in Canada, had little choice but to flee the onset of the White River volcanic eruption 1,300 years ago, an action that led to their permanent migration to the American Southwest, where their descendants became the Navaho and Apache. Natives of central and western El Salvador two thousand years ago were also forced to migrate after the eruption of Ilopango.[60]

Studies of natural disasters since the 1950s have shown that after a disaster people may experience "disaster syndrome" and be completely docile, while "some may experience 'counterdisaster syndrome' exhibiting a euphoric identification with the community, physical overexertion and low efficiency with an uncritical acceptance of 'leaders' who emerge during the rescue. . . . The effects may last from a few hours to many days."[61]

EFFECTS OF THE MINOAN ERUPTION

The physical and psychological suffering of the peoples of the Aegean who experienced the Minoan eruption must have been considerable. Various gods were undoubtedly invoked to rescue people from this bizarre series of catastrophes. In part of the Greek poet Hesiod's (c. 700 B.C.E.) poem *The Theogony* ("The Birth of the Gods") there is a battle between the gods of Olympus and their enemies, the Titans, that is clearly a recounting of the Santorini eruption, though no specific human suffering is recounted.[62] In the Hittite myth of Ullikummi the gods sever what is arguably the ash pillar produced by the Minoan eruption from the shoulder of the being who supports the world, thus saving it. In mainland Greece the goddess Athene is credited with saving the Acropolis of Athens from the effects of a tsunami that may have been generated by the eruption.[63]

In the Egyptian Delta, Seth, god of thunder and storms, would have been called upon by the Hyksos in the face of such a disaster. An incantation found in an Egyptian medical papyrus dated to the reign of Amenophis I (at about 1550 B.C.E.) refers to Seth having banned the Mediterranean Sea.[64] Avaris was about twenty-eight kilometers from the sea at that time, and a tsunami could not have reached that far. However, tsunami waves could have briefly caused some higher water levels in canals and river branches that fed in from the sea. Archaeological finds at Avaris include inscriptions dedicated to Seth by a later Hyksos ruler, Apophis, while a later Egyptian text states that Apophis made Seth his lord and did not serve any other god in the entire land except Seth, that he built a temple to

Seth, and made daily sacrifice to him.[65] The Hyksos must have credited the god Seth with something extraordinary to have given him such devotion.

Perhaps another group of western Semites living in the northeastern Delta region—say, in the Wadi Tumilat—likewise credited their god with some extraordinary acts at the time of the Santorini eruption.

CHAPTER FOUR

The Plagues, the Exodus, and Historical Reality

The account of the ten plagues and the Exodus from Egypt has fascinated both scholars and ordinary people for centuries. In fact, "plague" is a bit of a misnomer, since most of these events can be better described as "signs" and "wonders." Were they real, or were they the products of literary composition centuries after the date of the Exodus? Let us look, first, at the scholarly opinion on the plagues, then at the way oral historians would interpret them, and then, with the geological information on the Minoan eruption and its effects, and the historical/archaeological material from the previous two chapters, go on to see how well the stories of the plagues in Exodus chapters 7–10 fit what we think happened in the northeastern Egyptian Delta following the eruption of the Santorini volcano.

SCHOLARLY OPINION, ORALLY TRANSMITTED TRADITIONS, AND THE PLAGUES

Most modern scholars adhere to some variety of the Documentary or J, E, D, and P Hypothesis, which holds that the books of Genesis through Numbers contain three source documents: J or the Yahwist, reflecting the monarchy of the kingdom of Judah in the ninth century B.C.E., E, the Elohist, stemming from a northern Israelite kingdom source in the eighth century B.C.E., and P, a postexilic (sixth or fifth century B.C.E. or later) Priestly source, which added to these first two. The last source document was D, or Deuteronomy through 2 Kings, which came from the seventh century B.C.E. court of the Judean king Josiah. A later editor or redactor, R (or several of them), joined all these sources together into their present form.[1]

Scholars differ, however, on which plagues go with which Documentary Hypothesis source. One, John Van Seters, assigns all the plagues either to the Yahwist (J) or to the Priestly (P) source. He identifies as belonging to the Yahwist (1) the Nile turned to blood; (2) frogs; (3) flies [this word is translated as insects or gadflies by other authors]; (4) the pestilence of the livestock; (5) hail; (6) locusts; (7) the death of the firstborn. The other plagues he ascribes to the P source. Van Seters considers that "the whole [J] plague narrative is so consistent in its pattern and so uniform in its outlook that it must be the literary artistry of a single author, the Yahwist."[2] An earlier twentieth century scholar, Martin Noth, similarly assigned the plagues to the J and P sources but came to a vastly different conclusion: "the set of plague stories is not a well considered literary product but is derived from living oral tradition. . . ."[3] Other scholars such as Georg Fohrer and Brevard Childs attributed the plagues to J, E (the Elohist), and P.[4]

In a different vein, Moshe Greenberg looked at the symmetry in the narrative unit and saw three sets of three plague episodes each, ending with the plague of darkness but not including the deaths of the firstborn.[5] Dennis McCarthy found a chiastic—one of the plot types that signals orality—structure to the plagues, with what he called the last plague, the darkness, corresponding to what he designated as the first, the rod of Moses changing into a serpent. George Coats discovered the chiastic structure only in his recreation of the series of J plagues.[6] With all these differences of opinion, it is no wonder Roland de Vaux wrote that "an examination of the first nine plagues, without taking the tenth into account, reveals a very careful literary composition which in fact defies analysis by the methods of literary criticism."[7]

However you count them or whichever ones you include, the first nine plagues, while having a definite pattern and repeated motifs (especially Pharaoh's "hardened heart"), also contain logical inconsistencies and repetitions. If *all* the water was changed to blood by Aaron's rod (Exodus 7:20) how could there be water left for the Egyptian magicians to do the same thing (Exodus 7:22)? Did the blood or the dead fish (Exodus 7:18) poison the water? The cattle that died by pestilence in the fifth plague were resurrected to die of boils in the sixth plague and re-resurrected to die once more by hail in the seventh plague. Are the biting mosquitoes (or gnats or lice) in the third plague the same as the flies or gadflies in the fourth plague? Certainly the murrain of the cattle in the fifth plague is nearly duplicated by the afflictions to cattle and people in the sixth.

But logical inconsistencies and duplications are exactly what oral historians would expect to find within stories that have been transmitted orally for great lengths of time, particularly when they have been passed down through different groups and then combined. The common oral characteristic of exaggeration (sharpening) accounts for many of the most obvious inconsistencies: *all* the cattle, *all* the crops, *all* of Egypt. The various anachronisms in the stories are also to be expected in orally transmitted tradition.

THE PLAGUES COMPARED TO THE MINOAN ERUPTION OF SANTORINI

Now, to set the scene for our comparison: It is 1628 B.C.E. The Hyksos, Canaanites from southwestern Asia, rule the Delta region and other parts of the Nile Valley. Their ruler calls himself a "ruler of foreign lands"; he does not use the name Pharaoh. Avaris, in its D/3 stratum, is a populous city, the Hyksos capital. Other settlements of western Semites are found farther south and southeast, at Tell el-Yehudiyah and in the Wadi Tumilat. In the Wadi Tumilat the main settlements are at Tell er-Retabah and Tell el-Maskhuta; there are smaller hamlets in the wadi as well. Although also western Semites, the people of the Wadi Tumilat may be a distinct social or political subgroup.[8] They have been living in the wadi for several generations, long enough to absorb some of the Egyptian pottery styles and to develop their own styles from what were originally Syro-Canaanite pottery types.[9] They are pastoralists who spend the hot summer months with their flocks around the wadi's ponds and wells. The wadi's towns and hamlets are their winter homes—they plant cereal crops in November, six weeks after planting starts in the south of Egypt, and tend to the donkey caravanners who come through the wadi in the winter months.[10] It is now the very end of January or the very beginning of February, and the people are still in their winter homes. Soon they will start their barley harvest; their emmer wheat is still growing and will be harvested at the end of March or in early April.[11]

About this time the people of the Delta may have heard a rumbling noise, much like thunder. They were too far from Santorini to see the plinian eruption column, which reached an estimated thirty-six to thirty-eight kilometers in height. Because of the Earth's curvature, the column would have had to have been nearly fifty kilometers high to have been seen in the Egyptian Delta.[12] Perhaps some of the people of the Delta noticed a clattering or shattering of some of their pottery as a wave of air seemed to rush past. They probably thought it must be Seth/Baal, their storm god, and that a winter storm was approaching from the northwest, as such storms have always regularly done.

According to the Exodus account, the first plague, or sign, occurred when all the water of the Nile, its tributaries, its canals, and the water in all the ponds, was changed to blood. The fish in the Nile died, and the Nile reeked. Tsunamis from the second phase of the eruption would have reached the Egyptian Delta in less than an hour, with wave heights of seven to twelve meters.[13] Normal waves striking northeastern Egypt range in height from 0.40–0.75 meters in summer to 1.5–3.0 meters during winter storms.[14] The Santorini tsunamis, then, would have been three to four times higher than the highest waves usually experienced on the Egyptian coast and, considering the Delta's flat topography, probably caused extensive flooding of the coastal plain as the waves were channeled up nearshore channels and canals, possibly affecting some of the freshwater lakes and ponds as well. Many of the normal drinking water sources would have become contaminated, and the oxygen content of the water would be disturbed by the increased turbidity. This would have been enough to kill a good many of the freshwater fish.[15]

Was the water turned to a blood-red color, or is this merely a common folktale motif? The Sumerian goddess Inanna, for example, sent a series of plagues on people to punish a human who raped her. The first of her plagues was the turning of all the water to blood, so that people could not drink.[16] But an Egyptian text possibly dated to the Hyksos time period, the Admonitions of Ipwer, contains the lines: "Lo, the river is blood, As one drinks of it one shrinks from people and thirsts for water."[17]

Toxic dinoflagellates are the little one-celled organisms that cause the deadly algal blooms or "red tides" along coasts around the world. In the Mediterranean these dinoflagellates are found in the sea off the deltas of the major rivers, such as the Nile, carried in the tidal current that flows from west to east just offshore. They grow best in tropical and subtropical seas and in the rainy season, which in the Mediterranean is winter.[18] Recent research has also shown that iron from windborne, iron-oxide-bearing dust that falls into an ocean or sea can be taken up ("eaten") by a tiny organism called *Trichodesmium*, which then excretes great amounts of dissolved organic nitrogen into the ocean water. This nitrogen in turn spurs massive growth of toxic dinoflagellates and results in a red tide two to three months after the original dustfall.[19] The precursor ashfall (BO_0), occurring a few months before the main Minoan eruption, did have a significant iron oxide content, and sulfuric acid (from sulfate) on the surfaces of the ash particles would have caused the iron to become soluble enough in the sea water for the *Trichodesmium* to use.[20] Carried by winds to the salt waters off the Delta, the Santorini dust would have caused a red algal "bloom" there by the time of the main eruption. Because tsunamis extend through the entire water column to the seafloor, when they reached the sea waters off the Delta they would have carried any toxic bloom or red tide ashore. The toxins would have killed a good many fish, and any that survived would be subject to the acid rains and tephra fall that came later. The tephra itself, iron-rich and rose-colored, would also have caused the waters to redden when it was washed into the river by those same rains.

According to Exodus 7:25, seven days passed before the onset of the next plague, the swarming of the frogs onto the land. Time in oral traditions is often exaggerated, as it was here. This "week" may signal the interval between the arrival of the tsunami and the onset of the next series of disasters. The amphibian invasion would in fact have happened rather soon after the contamination of freshwater habitats from the debris and flooding caused by the tsunamis, followed by a massive die-off when the frogs stayed away from the water for too long.

Given the estimated range of wind speeds mentioned in the previous chapter, the first ash from the plinian eruption would have reached the Delta in eight to thirty-two hours, well after the noise of the eruption, the atmospheric shock wave, and the tsunami had come ashore. Only the finest ash particles would have been carried this far. In Exodus, the die-off of the frogs was followed by the plague of the gnats (or some other small biting insect). The gnats were produced when Aaron, Moses' brother, struck the dust of the ground with his staff or rod and the dust became gnats that landed on man and beast.

One common type of error, especially in group remembrance, is implicational, when people try to "make sense" of the story. Dust usually comes from the ground, and so it does in this present version of the story when, becoming transformed into gnats, it was the only way to make sense of biting dust. Originally, though, the biting dust came from the air in the first winds that carried fine ash from the initial stages of the plinian eruption cloud to the Delta. That it was not accompanied by water in some form suggests that at this stage the water in the eruption cloud was evaporating before it reached the ground.[21] This first light ashfall was not dense enough to produce darkness; it was only dense enough to be perceived as dust—an acid-bearing dust, irritating the skin of man and beast, like gnats or lice or mosquitoes biting. In time, the modifier "like" would be dropped from the oral tradition, as modifiers are in the leveling process, and the dust was transformed into small biting insects.

This acid dust was followed by swarms of insects coming into the houses of the people, except in the land of Goshen. Insects are particularly vulnerable to tephra fall, losing their surface wax layer and becoming dehydrated. Houseflies, yellowjacket wasps, and various sorts of bees lost much of their body moisture and died in the hours following the Mount St. Helens tephra fall in May, 1980. Ash also blocked their tracheal tubes and hindered their ability to fly.[22] Insects in the Nile Delta would have made some attempt to seek shelter when the tephra fall began, much as birds sought shelter in the houses of New Guineans during the mid-seventeenth-century Long Island tephra fall.[23] An alternative possibility is that the flies were simply an embellished version of the gnats, which grew bigger through retelling and eventually, when incorporated into a general version of the narrative, were included as a separate plague. That the land of Goshen—that is, the Wadi Tumilat—was said to have been free of insect swarms may have been a later theological and nationalistic insertion: "But on that day I will set apart the land of Goshen, where my people live . . . ; that you may know that I the Lord am in this land" (Exodus 8:22 (18 MT)).[24] As later generations tried to make sense of these stories, it would only have seemed right that the Egyptians, but not the Israelites, were affected by these and subsequent plagues.

According to Exodus 9:3–6, God then sent a pestilence onto the livestock *in the field*: the asses, the camels (an anachronism, camels came centuries later), the oxen, and the sheep, but not those of the Israelites. This pestilence would have been caused by the animals breathing the acidic dust as they stood in the fields (the people, presumably, would have fled inside) or by ingesting the fallen ash while browsing on near-ground vegetation, just as the reindeer were affected by the eruptions of Unimak (1825) and Katmai (1912) volcanoes in Alaska.[25]

The next wonder involved the tossing of furnace soot by Moses and Aaron into the air. As soon as the soot was tossed heavenward it became boil-blisters on the skins of humans and animals. The Hebrew word used is related to the Ugarit word "burn."[26] This again describes the association of airborne dust with irritations of the skin and suggests (starting with the plague of gnats and going on to the pestilence of the animals and that of blisters and skin irritations) a continuous, ever-worsening fall of ash on the northeastern Delta. As the various manifestations and worsening effects of the ashfall became drawn out and stylized in later retelling, these manifestations became discrete events, or "plagues." This is also true of the next three plagues.

After the skin irritations came the seventh plague, a violent hailstorm with thunder and fire: "the Lord sent thunder and hail, and fire came down on the earth . . . there was hail with fire flashing continually in the midst of it, such heavy hail as had never fallen on the land of Egypt" (Exodus 9:23, 24). The hail ruined many of the crops (Exodus 9:31–32), and was also associated with a heavy rainstorm (Exodus 9:33, 34). Meteorological turbulence or thunderstorms will enhance aggregation of particles in an eruption cloud, and so precipitate a secondary maxima ashfall at great distances from an eruption.[27] Turbulence from a cyclonic storm could easily have caused the icy ashballs from the electrically charged Santorini eruption cloud to aggregate and fall to earth over the Delta, to be perceived as hail shot through with lightning, followed by rain from the storm itself.

After the hail, fire, and rain came what would normally be an ordinary occurrence, a locust plague—locusts are expected in the Delta in the late winter or early spring.[28] In any case, the Exodus account says that God reversed an exceedingly strong sea wind, which blew the locusts into the Sea of Reeds. This accurately describes the counterclockwise rotation of winds on the southern edge of a cyclonic winter storm system coming in from the Mediterranean Sea.

Next came the ninth plague, of darkness, which lasted for three days in Egypt, except in the land of Goshen. The ash cloud appears to have reached its greatest density and extent at this point, covering the Wadi Tumilat along with the rest of the northeastern Delta. The darkness would have occurred with or immediately after the hail, but the recounting of it, plus the story of the locusts, would have been drawn out in oral recitation. How long the darkness lasted is an open question. Three days was the length most often attributed to the New Guinea time of darkness, but in both instances, it seems to have been an exaggeration caused by fear and disorientation. Richard Blong has calculated that each centimeter of uncompacted tephra on the ground will produce an average of 4.8 hours of darkness, but recorded values vary considerably.[29]

By now, all the people of the northeastern Nile Delta would have been extremely frightened. They were in the middle of a natural disaster, the like of which they had never seen before. Applying the general theory of human behavioral adjustments to natural disasters developed by Ian Burton and his coworkers,[30] the people would have already passed the first threshold (that of conscious awareness of the disaster) and also the second, in which active loss reduction measures are undertaken (such as staying indoors and sheltering their livestock). This is when religious measures are usually undertaken.

According to the Exodus account, the Egyptian magicians were called on to duplicate the water turned blood, the dead fish, and the frogs, but they failed to duplicate the plague of gnats and leave the story when they become covered with boils in the sixth plague. In reality, the Hyksos king would have attempted loss reduction by demanding that his magicians—or his priests—call upon some divinity to end the calamities, not duplicate them. Moses and Aaron had nothing to do with these measures, but were inserted into the story at a later date. The ruler, who was ultimately responsible for the well-being of his people and harmony with the gods, had to do *something*, or he would probably be supplanted—a later Hyksos ruler, Apophis, may have been a usurper.[31] The Hyksos' veneration of the god Seth suggests that they directed their sacrifices to him, probably by sacrificing donkeys, Seth's animal, much as the New Guineans sacrificed their pigs.[32] The incantation found in the Egyptian medical papyrus mentioned in the previous chapter suggests that the people believed Seth came to their aid.

While the Hyksos ruler and his priests were sacrificing to Seth, some of the ordinary people of Avaris would have passed the third disaster threshold, that of intolerance. Radical action needed to be taken: "Such radical action involves in situ fundamental adaptive changes, or, in extreme cases where the environmental changes are beyond the human technological capacity to cope, migration occurs."[33] In short, people flee, either temporarily or permanently. No doubt a good many of the people of Avaris feared death if they stayed in their city where the burning rain, the hail, the tephra cloud, and its darkness hung over them. Some would have fled south to the Wadi Tumilat.

THE TWO EXODUSES: FLIGHT AND EXPULSION

At this point we encounter a set of inconsistencies in the Exodus story that goes to the very heart of the narrative. This problem is most clearly set out by Roland de Vaux: "There are, in fact, two distinct presentations of the exodus story, the exodus-flight and the exodus-expulsion."[34] These two stories were joined together orally centuries before the tradition came to be written down—"merged together at a very early stage and . . . had a deep influence on each other."[35] Scarcely anything remains of the older version of the story, the one that includes Moses. Although Moses is present in the younger story, he is a later insertion. Instead, Exodus 5:3, 5–19 clearly indicates that in the younger story the representatives or elders of the Israelites, not Moses and Aaron, do the negotiating with Pharaoh, a term that did not come into use until Egypt's Eighteenth Dynasty.[36] Here the elders ask permission to go on a three days' journey into the wilderness to make slaughter-offering to the god of the Hebrews. Exodus 7:16 and 8:8 (8:4 MT), 25–28 (21–24 MT) also mention this offering, while Exodus 10:9 mentions having a feast for the Lord. These passages make it clear that this sacrifice is already an established custom among the Israelites.

In the younger story, Pharaoh refuses to let the Israelites go and breaks off negotiations in Exodus 10:28. Closely linked to these negotiations is Exodus 8:26 (22 MT), which contains a time-marker. Moses (actually the elders or representatives of Israel) tells Pharaoh that the sacrifices they offer to God would be offensive (an abomination) to the Egyptians and that the Egyptians might stone them for it. This passage indicates a post-Hyksos date, when the native Egyptians were once again rulers of all Egypt, for only native Egyptians would be offended by mass slaughter sacrifices of sheep or rams, sacred to the Egyptian god Amon-Re.[37] Other Semites, like the Israelites themselves, had a long religious tradition of sheep and goat sacrifice, as the archaeological remains from Avaris and many other sites confirm.[38] This post-Hyksos version is linked with the tenth plague, the death of the firstborn (Exodus 12:29–30) and to the exodus-expulsion—Pharaoh drives the Israelites out (Exodus 6:1*b*; 11:1; 12:31–32). It will be discussed in a later chapter of this book.

The older story, the exodus-flight, is the one linked to the first nine plagues. One fragment referring to this exodus is preserved in Exodus 14:5*a*: "When the king of Egypt was told that the people had fled . . ." A larger fragment is found in Exodus 12:33: "The Egyptians urged the people [of Israel] to hasten their departure from the land, for they said, 'We shall all be dead.'" This verse makes a great deal of sense if the "Egyptians" were in fact panicked Avarans fleeing south to the Wadi Tumilat. The people of the wadi probably did not need much urging to flee for their own lives, as recounted in Exodus 12:33, for they were frightened by the unknown darkness and the other catastrophes they were experiencing.

THE EXODUS AND THE UNLEAVENED BREAD

This crucial narrative information in Exodus 12:33 is followed by the story of the *matzoh* in Exodus 12:34: Dough has been put in kneading bowls to ferment naturally. When the Israelites flee they wrap up their kneading bowls and carry them on their shoulders, but the dough has not had time to acquire wild yeast from the air and is baked into unleavened bread, matzoh. As Greta Hort observed, it is the making and eating of the matzoh that actually commemorates the exodus from Egypt.[39] It has long been suggested that the eating of the matzoh had its origin in the ancient Canaanite Feast of the Unleavened Bread, a harvest festival. However, the month of Abib (March–April) is not a time of harvest in Canaan, nor does the Hebrew Feast of Unleavened Bread reflect agricultural activity.[40]

The term "kneading bowls" in Exodus 12:34 is sometimes translated as "kneading troughs," but "kneading bowls" perfectly describes many of the broad shallow woklike platter-bowls found at Tell el-Maskhuta. Although kneading dough in such a vessel would not work very well on a flat surface, putting the platter-bowl in a scooped-out depression in the sand or earth to anchor it and kneading on one's knees works quite well (I experimented with a similar-shaped platter). The bowl shape has one great advantage over a flat trough—the precious flour, ground so laboriously by hand, doesn't escape and thus isn't wasted. Carried in a platter-bowl within a cloak hitched across the shoulder (on the back, really), the dough would stay put; in a trough it would fall out.

The method of bread making suggested by Exodus 12:34 is quite unlike that used by the ancient Egyptians. As early as 2400 B.C.E. Egyptians were producing cone-shaped, leavened bread from barley and emmer wheat, allowing the dough to ferment in large vats and kneading it with their feet in vats or large wooden troughs. Later, bread was produced in a variety of shapes, from flat pitalike to rolled loaves of barley to high cone forms. The Egyptians did not have high-gluten bread wheat (*Triticum aestivum*) until the second half of the first millennium B.C.E. but had developed almost pure domestic bread yeast by 1500–1450 B.C.E., not too long before large centralized bakeries appeared.[41] Foreign workers or slaves living in Egyptian towns, especially in the New Kingdom, would not have made their own bread, but rather would have had their bread issued to them as rations.[42] But autonomous groups of Semites living in the Delta during the Hyksos era would probably have made their bread as described in Exodus 12.

FLIGHT FROM THE WADI

A panicked horde, the people of the wadi and the Avaran refugees fled eastward past Succoth—Tell el-Maskhuta and its environs—and then out the eastern end of the wadi. The Avarans who accompanied the Israelites were the "mixed multitude" of Exodus 12:38. The term is better translated as "riffraff."[43] I wonder if it originally conveyed the idea of "refugees." It was these people who carried the vivid testimony of what had happened in Avaris, a set of memories that became fused with those of the people of the Wadi Tumilat into the "standard version" of the signs and wonders that made its way into Israelite tradition as the first nine Exodus plagues.

The route of travel mentioned in Exodus 13:20 and Numbers 33:6—Succoth to Etham—belongs to this original exodus; the sequence that includes turning back to Pi-hahiroth and camping by Baal-Zephon and before Migdol (Numbers 33:7b–8a) refers to the later exodus-expulsion. In this first exodus, after leaving Succoth the people journeyed to Etham: "They set out from Etham, . . . passed through the sea into the wilderness, went a three days' journey in the wilderness of Etham and camped at Marah" (Numbers 33:7a, 8b).

Once outside of the wadi the people fled south, around the western edge of what is now Lake Timsah toward the Bitter Lakes. Both Lake Timsah and the Bitter Lakes were probably included in the Egyptian term *km wr*, and it is possible that an ancient frontier canal at least partially connected them.[44] South of Lake Timsah there are two ridges that could have allowed the people of the wadi to have crossed over to the Sinai Peninsula: one connects the Great and Little Bitter Lakes, while the second occurs about twelve kilometers south of the present town of Suez in what is now the northernmost extension of the Gulf of Suez (see arrows, figure 2.1).[45] Under the right conditions, notably after the steady blowing of a strong wind for several hours, either of these ridges would have been exposed to the air.[46] Dryshod, the people of the wadi and the Avaran refugees, and their animals, would have crossed into the Sinai Peninsula. Now that the disaster was past, they were ripe for the "counterdisaster syndrome" mentioned in the previous chapter. For a short time at least, they gave their uncritical acceptance to the leader who had emerged during the crisis, a leader who was now demanding that they journey through the wilderness to a mountain where he had spoken to God.

His name was Moses.

CHAPTER FIVE

Moses and the Mountain of God

Who was Moses? Without any doubt, he is the key human figure in the Exodus story. Without him there would have been no Exodus, no journey to the holy mountain, no sojourn in the wilderness, and no return of the people of Israel to Canaan. The first five books of the Bible are traditionally attributed to him, and the Ten Commandments, given to him on the Mountain of God, are arguably the fundamental religious, legal, and ethical guidelines for Western Civilization. One would think that there is nothing new to be said about him, but by putting Moses into the historical context of the Hyksos occupation of Egypt new insights about this seminal individual appear.

MOSES AND THE FAMILY OF LEVI IN EGYPT

Moses is an Egyptian name, or at least the second half of one. It means "the god [. . .] is born" and was usually given to a child who was born on the birthday of a particular god. The account in Exodus 2:10 gives a Hebrew meaning to the name, claiming it means "to draw out."[1] The term "Hebrew" (*'ibrî*) in Exodus 1:15–22 is related to the term "Habiru" or "'Apiru," which is found in hundreds of Near Eastern texts in the second millennium B.C.E. From these texts, we know that Habiru were bands of uprooted people—migrants—usually followers of a prominent leader who moved into a new area and lived as foreigners or aliens under the local ruler. These people often had a military role. For example, the eighteenth century B.C.E. Mari texts (Mari is in northern Mesopotamia) describe a Habiru leader and his troops.[2]

The term Habiru aptly describes Jacob and his familial band migrating to Egypt from north-central Canaan during a time of famine, as Abraham had in an earlier time. One famine, during the reign of Thirteenth Dynasty king Sobekhotep III (ca. 1749–1742 B.C.E.—see table 2.2), may have brought numbers of Asiatics to Egypt, with Memphis serving as a clearinghouse for them. A list of domestic servants (or slaves) from the reign of Sobekhotep III includes at least forty-eight names (out of an original ninety-five) of northwestern Semitic origin. One is a close approximation of the name of the midwife in Exodus 1:15, "Shiprah."[3]

A fragment from the late third or early second century B.C.E. Hellenistic writer Artapanus (who wrote a work called "On the Jews") says that Chenephres, who was king over the regions beyond (south of) Memphis, married Merris, a daughter of a northern Egyptian king named Palmanothes, there being many kings of Egypt at that time. Because she was barren Merris adopted a child of the Jews and named it Moses.[4]

Artapanus includes a number of folk traditions that clearly have been passed down orally for some time,[5] but this particular story is interesting because it does not agree with the biblical account of Moses' adoption in Exodus 2:5–10, where the Egyptian princess is referred to as Pharaoh's daughter. In contrast to the biblical tradition, Artapanus's tradition remembered Merris's Egyptian connection (wife of the Pharaoh Chenephres) and the Egyptian names. There is also no mention of a baby in a basket retrieved from the reeds. These are indications that Artapanus is recounting an Egyptian tradition independent of the biblical account. The name of the Egyptian ruler in Artapanus's story, Chenephres, is Khaneferre Sobekhotep IV (ca. 1732–1720 B.C.E.). Egyptians generally knew their rulers by their prenomens, such as Khaneferre, not Sobekhotep. Khaneferre Sobekhotep IV's capitol was just north of Memphis, and at this time independent rulers were establishing (or had established) themselves both to the north, in the Delta, and to the south, in Thebes and in Nubia. He also had more than one wife.[6]

Khaneferre Sobekhotep IV was a contemporary of the beginning rulers of the Asiatic Fourteenth Dynasty in Avaris, Nehesy and his father (see chapter 2). A dynastic match with Nehesy's sister or daughter would have been a sound political move by an Egyptian ruler in Memphis. I think the name Palmanothes might be a garbled form of Ptah-moses, "the god Ptah is born." Ptah was the patron god of Memphis and, in Middle Kingdom times, many Egyptian names for Asiatics were compounded with "Ptah."[7] Rather than being the name of Merris' father, Palmanothes (or Ptah-moses) was more likely the name of the son Merris adopted, since there is only enough space on the Turin king list for a very short name for Nehe-sy's father, not a long one such as Palmanothes or Ptah-moses.[8]

A biblical story that probably relates to this dynastic alliance is found in Genesis. Genesis 12:10–13:1 tells how Abraham and his wife Sara go to Egypt in a time of famine. In verses 14–16 Sara is taken into Pharaoh's house. Variants of this story appear in Genesis 20, where the king who takes in Sara is Abimelech, King of Gerar, and Genesis 26, where instead of Abraham it is Isaac going to Gerar in another time of famine, the woman is Rebekah his wife and ostensible sister, and the king is Abime-lech, King of the Philistines. This story has obviously become duplicated and altered through time, but at its core is an account of nomads who have gone somewhere to seek relief from famine and of a sister or wife being taken into the harem of a king when they arrive.

Rather than being from the time of Abraham or Isaac, I think this story relates to the time of Jacob and his son Levi and their family, mi-grating to Egypt because of a famine in Canaan, as recounted in Genesis 45. Egyptian names appear in the family of Levi: Assir, son of Korah, Levi's grandson; Moses, son of Amram, another of Levi's grandsons; Merari, Levi's youngest son (who in turn also had a son Moses); and Phinehas, son of Eleazar, son of Aaron (see figure 5.1). Merari, like Mer-ris, is *mrry*, a common Middle Kingdom name meaning "beloved." The name of Aaron's sister Miriam also features "beloved," in this case: "be-loved of Ya[weh]." Assir is Osiris in Egyptian. Phinehas is *P'-nhsy* or Pi-Nehesy, a Late Egyptian form of Nehesy, "the Nubian."[9] This name is also found as a place name in the Delta near Daphnae (Tell Defenneh): "the place of those of the Asiatic Pi-Nehesy." Several scholars believe this Late Egyptian place name refers back to the Asiatic Fourteenth Dynasty king, Nehesy.[10] Nehesy is the only individual known as both an Asiatic and a Nubian.

Genealogies in the Hebrew Bible are important because they express "all sorts of social, political, and religious relationships," and often change in various versions.[11] The genealogy of Moses in Exodus 6:16, 18, and 20 (and Numbers 26:59) states that Amram, son of Kohath, married his aunt, Jochebed, the daughter of Levi. This most likely represents the fusion of two genealogies. If we assume that Jochebed was one of the very first of Levi's children, born, say, about 1740 B.C.E., she could have gone with the Fourteenth Dynasty princess Merris to the harem of Khaneferre Sobekhotep IV and there in about 1725–1720 B.C.E. had a son named Ptah-moses who was adopted by Merris. Merris would not have been

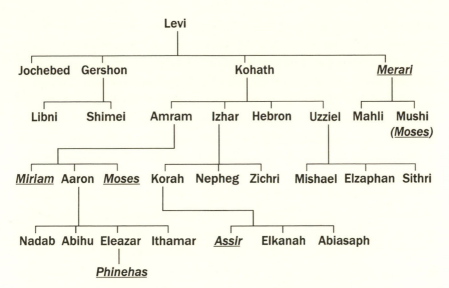

FIGURE 5.1. The descendants of Levi, as recounted in Exodus 6:16–25 and Numbers 26:57–60. Names of apparent Egyptian origin are italicized and underlined.

permitted to adopt just any child, so this infant was almost certainly Khaneferre Sobekhotep's son, in the same way that Hagar had a son, Ishmael, for Abraham. Such an adoption would have cemented the alliance between the Thirteenth and Fourteenth Dynasties and brought honor to both the family of Nehesy and that of Levi. Genesis 12:16 says "and for her sake he dealt well with Abram"; the original story probably had the name Levi rather than Abram. The biblical Levites gave their children names to commemorate this honor, names that parallel some of those in the story in Artapanus.

In later centuries, when Jochebed's original role became incompatible with Israelite tribal or nationalistic feelings, the story of the baby in the basket was adapted from the birth tale of Sargon the Great of Akkad in Mesopotamia.[12] In this tale, Sargon, king of Akkad in the second half of the third millennium B.C.E., was the son of a high priestess. Since priestesses were not supposed to have children, his mother put her infant son in a woven basket caulked with pitch and cast it onto the river. The basket and baby were found by Aqqi, a water drawer. Aqqi means "I drew out," an identical meaning to that for the name Moses given in Exodus 2:10.[13] The accretion of the Sargon birth story onto the story of Jochebed's son Moses deprived the original story (i.e., that of a Levite woman in an Egyptian royal harem having a son) of its true meaning and context, and consequently only a displaced fragment survived in the Genesis stories of Sara and Rebekah. Here is a good example of how oral traditions are reinterpreted and modified as the needs and aspirations of a people change through time.

The Moses adoption story is clearly too early for the biblical Moses, son of Amram and leader of the people from the Wadi Tumilat in the later seventeenth century B.C.E. This is what oral historians call a "descending anachronism," when an event is moved from an earlier to a later epoch.[14] Such anachronisms occur when founders or culture heroes (such as the biblical Moses) are credited in oral tradition for events or accomplishments of earlier figures. The Moses of the Exodus would then have been a relative as well as namesake of the adopted royal Moses, good reasons for him to be associated with the latter in later oral tradition.

AN ARCHAEOLOGICAL INDICATION OF THE MOSES ADOPTION STORY?

In one of the unphased tombs at Tell el-Maskhuta, a young woman seventeen to thirty years of age and a child about 6.5 to eight years of age were buried with grave goods that included a Khaneferre Sobekhotep IV scarab. This tomb had a number of relatively rich grave offerings: a silver choker necklace, a number of silver and bronze earrings and toggle pins, amulets of faience and steatite with the child (faience is a ceramic made of ground quartz), faience, carnelian, and amethyst beads, a necklace of amethyst, gold, and faience beads, three design scarabs, a steatite cylinder seal, and three cups, six juglets, and a ringstand, along with an offering of sheep or goat remains. There may originally have been many more valuables, for the tomb had been broken into in antiquity.[15] Vaulted tombs were most common in the earlier occupation layers of Tell el-Maskhuta, and adults were not buried at the site in the later phases, two good reasons for placing this particular tomb early in the occupation sequence.

It seems likely that Jochebed was returned to her people after her son Moses was weaned (cf. Exodus 2:7–10) and that she would have been given gifts such as these, including the king's scarab, as a reward for her contribution to the alliance between the Thirteenth and Fourteenth Dynasties. These gifts would have signaled her own family's honor as well. Jochebed may well be the young woman in this tomb, perhaps dying in a later childbirth after marriage to one of her own people. The child buried with her may have been another offspring of hers.

Exodus 1:8–10 says: "Now a new king arose over Egypt, who did not know Joseph. He said to his people, 'Look, the Israelite people are more numerous and more powerful than we. Come, let us deal shrewdly with them or they will increase and, in the event of war, join our enemies and fight against us and escape from the land.'" Leaving aside "and escape from the land," and replacing "Joseph" with "Jacob," this is an accurate description of the political picture in the Delta at the onset of Hyksos rule. When the Hyksos under Salitis (Sheshi—see chapter 2) and his underkings took possession of the Delta they gained control of large numbers of other Semites already living there, as evidence from the earlier strata (G through E/1) at Tell el-Dab'a and from other Delta sites clearly shows. The early Hyksos may have viewed these other Semitic peoples as potential threats who could join together and take control of the area for themselves or form an alliance with a native Egyptian ruler and likewise displace the Hyksos.

With the coming of the Fifteenth Hyksos Dynasty to the Delta the people of the Wadi Tumilat would have been made to acknowledge Salitis/ Sheshi as their ruler and co-opted into taking care of the incoming trade caravans at the cafeteria-style facilities found in the later levels at Tell el-Maskhuta (see chapter 2). These facilities were probably the "flesh pots" in the land of Egypt referred to in Exodus 16:3. Also, a Sheshi scarab was found in one of the wadi's later tombs. But there was tension between the people of the wadi and their Hyksos overlords. Exodus 2:11–15 contains the story that Moses, while visiting the enslaved Hebrews, killed an Egyptian who was beating one of his kinfolk. Other Hebrews soon knew of this act and Pharaoh, when he heard of it, sought to have Moses killed.

This story makes little sense if Moses were truly the adopted son of Pharaoh's daughter (surely he could have ordered the Egyptian to stop beating the man), but is quite understandable in the context of a Hyksos ("Egyptian") official beating a man of the Wadi Tumilat and having one of this man's kinfolk (that is, Moses) defend his relative by killing the official. The Hyksos king would plausibly then attempt to have Moses killed.

Because of this homicide, Moses had to flee into the desert, crossing the Sinai Peninsula and passing into the country of Midian. There he met a priest of Midian, married the priest's daughter, and remained for many years. One day Moses took his father-in-law's flock "beyond the wilderness, and came to Horeb, the Mountain of God. There the angel of the Lord appeared to him in a flame of fire out of a bush; he looked, and the bush was blazing, yet it was not consumed" (Exodus 3:1–2). Moses had what scholars call a theophany, a meeting with God.

MIDIAN: BIBLICAL AND MODERN

The biblical Midian is rather poorly defined, but it appears to be south and east of Canaan and east of the Sinai Peninsula (see figure 5.2). In the book of Genesis (25:1–4) Midian is said to be the son of Abraham by his concubine Keturah. Keturah (*q^etûrah*) means "frankincense," a valuable aromatic which comes from the bark of trees that grow principally in southern Arabia. Midian's five sons Ephah, Epher, Hanoch, Abida, and Eldaah (Genesis 25:4) were actually desert oases in northwest Arabia inhabited by individual tribal groups. Abida corresponds to al-Badʿ, an oasis that became the second station on the Muslim pilgrim road from the Gulf of Aqaba to the Islamic holy cities. Ephah is probably Rwafa or Rawafa, near the north end of Harrat ar Raha in northwest Arabia.[16] Since the Midianites were northern Arabian tribes, but connected with the South Arabian frankincense, they were probably the donkey caravanners who brought the frankincense north.[17] Later, in the book of Judges, the Midianites had camels (domestic camels appear at the end of the second millennium B.C.E.) and were described as coming from east of the Jordan River (Judges 6).

What is known as Midian (or Midyan) in modern times is an area of northwestern Arabia beginning east of the Gulf of Aqaba (the Midian Peninsula) and extending along the Red Sea to Al-Wajh (26° 10′ North latitude)—see figure 5.2. The oasis of al-Badʿ (28° 28′ North latitude) on the Midian Peninsula is usually thought to be the town of Midian. Greta Hort, however, meticulously traced the history of the name "Midian" and discovered several Midians. The most prominent of them was a well-known town near the Arabian coast slightly north of about 26° 45′ North latitude. As early as the first century of the Common Era, this more southerly Midian had a traditional connection with Moses. The modern town of al-Badʿ only acquired the Moses tradition after the tenth century, when the resident tribe lost the site of the southerly Midian to another tribe.[18]

MIDIAN, PLATE TECTONICS, AND THE ARABIAN VOLCANOES

Geologically, Midian is on the border of two tectonic plates: the African and the Arabian, and a subplate that constitutes the Sinai Peninsula. The Arabian plate has been rotating away from its African neighbor for millions

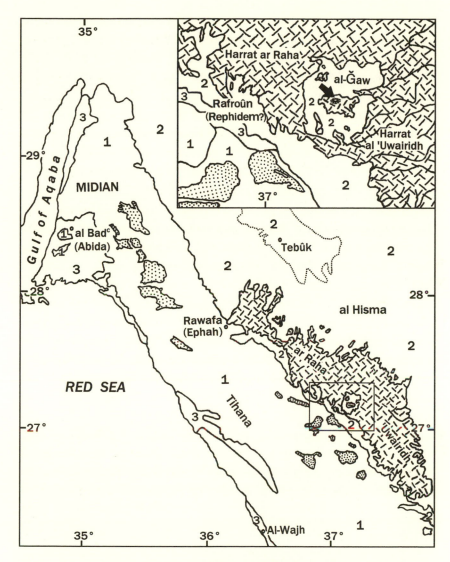

FIGURE 5.2. Simplified geological map of Midian with the Harrat ar Raha and Harrat al 'Uwairidh. Inset (see box in main map) shows the al-Ğaw basin with the volcano Hala'-l-Badr indicated by the arrow. Woven-lined areas mark lava flows. Dotted line around Tebûk marks the outer edge of the Tebûk basin. 1 = Precambrian age metamorphic and igneous basement rocks, with large-scale granite intrusives shown by dotted areas. 2 = sandstones. 3 = Tertiary and Quaternary age loose sands and gravels.

of years, with the area to the north of the Gulf of Aqaba as its hinge. This rotation has opened up the Red Sea, the Gulf of Suez, and, most recently, the Gulf of Aqaba. It has also stretched the earth's crust and thinned it, while hot material from the asthenosphere deep under the crust has pushed upward, creating a subterranean dome (the Afro-Arabian dome) extending from Ethiopia in the south to the Dead Sea Rift in the north.[19] The broad northern crest of this dome (the West Arabian swell) runs north-south through northwestern Arabia; along it are a series of cinder cones and fissures that have produced about 180,000 square kilometers of alkaline basalts, the Arabian harrat volcanoes.[20]

These Arabian volcanoes are different from the Santorini volcano. Unlike the magma typically erupted in subduction-zone volcanoes such as Santorini, the basaltic magma of the Arabian harrats is chemically less explosive than the Santorini/Thera magma, so it often flows effusively out onto the earth's surface as waves of hot black lava, transforming large stretches of northwestern Arabia into a vast, black wasteland. Chemically, the lava is not unlike some of the lava that comes out of the volcanoes in Hawaii.[21] Sometimes the lava forms cinder cones and craters, from which come columns of smoke and fire. Other times outpourings of lava are hurled into the air by their own expanding gases and form fire fountains.

Visitors to Kilauea Volcano in Hawaii are familiar with the black, lava-covered landscape of a basaltic volcano. They may even have seen a fire fountain, or a picture of one. Fire fountains occur when the gas trapped in basaltic lava escapes into the air, taking droplets of the lava with it, just like liquid shot from a spray bottle. These jets of incandescent liquid rock can shoot hundreds of meters into the air, or they may reach only a meter or so in height. They often spray for hours, swelling, dying away, and surging up irregularly.[22] This sort of fire show could certainly resemble a burning bush, especially from a distance, with real bushes silhouetted against the red glow of the molten lava shooting up into the air (see figure 5.3).

MOSES' RETURN TO EGYPT

Sometime after his encounter with God at the burning bush, Moses learned that those who sought to have him killed had died, and he returned to Egypt. Once back, Moses wanted his people to depart from Egypt and worship at the mountain where he had encountered the god of their ancestor, Abraham, and then return to Canaan. Anthropologists would label his efforts a nativistic or revitalization movement; sociologists commonly refer to it as a crisis cult. These movements have been common throughout human history and arise under conditions of hardship, such as political subordination or economic distress (or both). As one scholar noted: "shattering change is often needed to bring them about.... Revolution,

FIGURE 5.3. U.S. Geological Survey photo of a lava eruption (fire fountain) by Don Swanson of an eruption at Mauna Ulu in Hawaii in 1969.

war, natural catastrophes, economic dislocation, contact with a seemingly invincible and imperialist foreign people: these are common catalysts."[23] The Hyksos overlords would have appeared as an invincible and imperialist people to the pastoralists of the Wadi Tumilat.

What often happens is that someone in the subordinate or distressed group has a personality-transforming dream or vision and becomes a charismatic leader. In many cases he or she seeks to revitalize the group with particularly important cultural or religious elements from the group's past.

Occasionally this revitalization also involves migration. In the sixteenth century, Tupi-Guarani tribes of Brazil followed their prophets on journeys in vain attempts to find a Land-Without-Evil. Earlier, at the end of the eleventh century, great numbers of European peasants followed Peter the Hermit and similar prophets on crusades to the Holy Land, only to be massacred by Turkish bowman in Anatolia or by angry Hungarians on the Danube.[24] Later, in the nineteenth century, thousands of devout Mormons followed Brigham Young across the Great American Desert to settle on the shores of the Great Salt Lake and found the state of Utah.

Few, if any, such charismatic visionaries or prophets are lucky enough to have a volcanic eruption come along to help persuade their group to follow them, but Moses must have been able to convince his people that the disastrous effects of the Santorini eruption were the work of the god he represented, and that this god wanted them to follow Moses to the mountain.

LOCATION OF THE MOUNTAIN OF GOD

The route of the Israelites to the Mountain of God has been subject to a good deal of controversy, partly because the location of the mountain has been in doubt. In the Exodus led by Moses from Succoth to Etham, the Israelites probably took the most direct route across the limestone shield of Sinai (the Way of Seir) heading for the northern tip of the Gulf of Aqaba (see figure 5.4). This route would involve some hardship, since there were few wells and water sources on the way, and indeed, after three days traversing this wilderness, the people complained to Moses that they had no water (Exodus 15:22b–25). At Marah (which means bitter) Moses threw a log into the water and it became sweet. From there they went to Elim, probably near the tip of the Gulf of Aqaba, where there was abundant water.

Since the fourth century of the Common Era, Christians have usually believed that the Mountain of God lay in the southern part of the Sinai Peninsula, being either the Mountain of Moses or Mount St. Catherine. Josephus, the first century Jewish historian (and a native of Jerusalem) variously placed Mount Sinai east of the Gulf of Aqaba or between Egypt and Arabia.[25] In an older Jewish tradition, a third century B.C.E. Egyptian Jew named Demetrius said that Moses went to Arabia when he went to Midian.[26] There is also a specific reference to the Mountain of God in a letter written in the first century C.E. by a Jew of the Diaspora. In this letter, preserved in the Christian New Testament, Saul of Tarsus—Saint Paul— wrote (Galatians 4:24–25): "Now this is an allegory: these women are two covenants. One woman, in fact, is Hagar, from Mount Sinai, bearing children for slavery. Now Hagar is Mount Sinai in Arabia. . . ." Saul of Tarsus was educated in a rabbinical school in Jerusalem where he was "zealous for the traditions of my ancestors" (Galatians 1:14). In his zeal he persecuted early Christians in Jerusalem, then headed for Damascus to continue his activities there. On the road, however, he received a vision. His account, in Galatians 1:16–19, differs somewhat from the version in Acts 9:1–27, composed by another writer about thirty years after Saul (now Paul) penned Galatians. By his own account, after he received his vision Saul went away at once into Arabia, and afterward returned to Damascus.

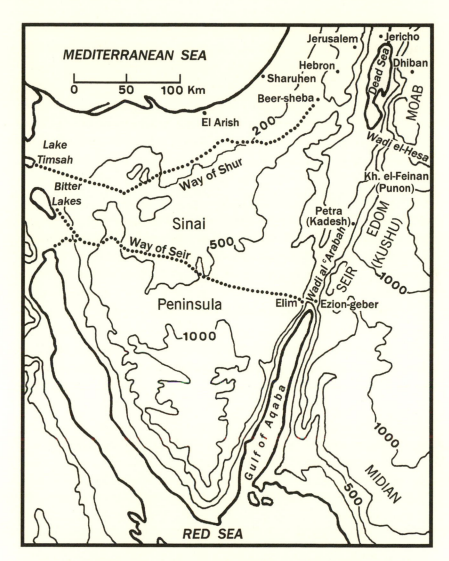

FIGURE 5.4. Map of the Sinai, northwest Arabia (Midian), and Canaan showing the route of the original exodus (starting at either the Bitter Lakes or the northernmost extension of the Red Sea and proceeding along the Way of Seir) to Elim at the Gulf of Aqaba, and the route of the later exodus along the Way of Shur back to Canaan. The contour lines are in meters. Ezion-geber, Kadesh, Punon and Dhiban were stopping places on the Israelites' return trek to Canaan at the end of the sojourn in the wilderness.

Now why should he go to Arabia? 1 Kings 19:8–18 describes how the ninth century B.C.E. northern Israelite prophet Elijah went to Horeb, the Mountain of God, and experienced a theophany there. The only reasonable explanation for Saul's journey to Arabia was that, like Elijah, he went to the Mountain of God seeking divine guidance. Saul must have had some idea of where the mountain actually was, and his reference to Mount Sinai in the Hagar allegory implies that he had located it in his travels.

Since the nineteenth century, some scholars have maintained that the Mountain of God was in fact a volcano in Arabia.[27] An eminent twentieth century scholar, Martin Noth, suggested that the Mountain of God would be found south of the oasis of *Tebūk*.[28] *Tebūk*, or Tebûk, lies in a basin just to the north of the two northernmost Arabian harrats: Harrat ar Raha and Harrat al 'Uwairidh, and is separated from them by a desolate sandstone plateau, al Hisma, that stretches from the granite mountains in northernmost Arabia and Jordan south toward the town of Medain Salah. There is an inland route from Aqaba through *Tebūk* to the holy city of Medina as well as a coastal route southward along the Red Sea. These two routes are separated by the two harrats, but one trail connects them just north of Harrat ar Raha and another passes through the broken area where the two harrats join.[29]

In 1910, Czech geographer Alois Musil passed through this area, mapping it and keeping careful records of his travels. On July 2 he traversed the al-Ğaw (or Al Jaww) basin, which partially separates harrats ar Raha and al 'Uwairidh. Here is what he saw:

> The valley broadens out into a basin enclosed on all sides by low, but steep, slopes, and known as al-Ğaw (the watering place) because it contains many *mšâše*, or rain water wells. The plain is covered with a fairly deep layer of clay in which various plants thrive luxuriantly, and it therefore forms the best winter encampment of the Beli [Bedouin]. The guide proudly pointed out to us the abundant withered pasturage through which we were passing and asked whether throughout our journey from Tebûk we had seen so many and such various plants. The annuals were yellowish, while the shrubs were a brilliant green. . . . Upon the eastern slope of the gray table mountain of Tadra is situated the black volcano Hala'-l-Bedr. On the western slope there used to flow a spring now said to have been clogged up by the collapse of a rock. . . . To the southeast we perceived the hill of Slej' and still farther in that direction the volcano of al-'Asi, in

which are the Morâjer 'Abîd Mûsa, "the caves of the servants of Moses." Our guide explained that servants of Moses sojourned in them when their master was abiding with Allâh. Another sacred spot is situated by the well of al-Hzêr. It is called al-Manhal, and upon it are 12 stones known as al-Madbah, where the Beli still offer up sacrifices when they are encamped close by.[30]

The volcano Hala'-l-Bedr (or Hallat al Badr, 27° 15' North latitude, 37° 12' East longitude), a cinder cone atop an expanse of flat, sandstone tableland (figure 5.2, inset), was the site of the most recent volcanism in the two harrats, having "vomited fire and stones" that killed many Bedouin and their camels and sheep, possibly in 640 C.E. Musil originally believed this was the site of the Mountain of God; later he changed his mind because he decided that the sacred mountain should be near the town of Midian, which he believed was al-Bad' on the Midian Peninsula.[31]

Another traveler to Arabia, Hermann von Wissmann, had an alternate candidate for the Mountain of God, a volcanic center along the western border of Harrat ar-Raha near the temple ruins of er-Rawafa.[32] This area is not far from one of the cross-trails that joins the inland and the coastal routes. It does not seem to have any local traditions connecting it with Moses, however, although it appears to be connected with the Midianite tribe of Ephah.[33] Either of these two proposed locations would fit Martin Noth's suggestion that the mountain was south of *Tebūk*, but Hala'-l-Bedr probably has more water in its immediate vicinity, and water is the most important determinant in Arabia.

A third Arabian traveler, Harry St. John Philby, wrote that the al Hisma plateau had a penetrating cold in the winter, and chilling winds. In that season the Bedouin left the Hisma highlands with their herds and moved to lower elevations, either to the *Tebūk* basin or south, to the Tihana coastal plain.[34] Going toward the coast in the fall and back to the Hisma plateau in the spring, the nomads had to pass north of Harrat ar-Raha or cross between the harrats via the al-Ğaw basin. They have undoubtedly followed this seasonal migration for thousands of years. In this way Moses could have passed by with his father-in-law's flocks and seen a fire fountain somewhere in the Harrat ar-Raha.

JOURNEY TO THE MOUNTAIN OF GOD

One feature that sustained the Israelites on their initial journey to the Mountain of God was a pillar of cloud and fire that went before them

(Exodus 13:21–22). This pillar appears on numerous occasions in the Exodus story: guiding the Israelites out of Egypt, at the sea between them and Pharaoh's army, in the wilderness, and often when God descends to speak with Moses in his tent. But as one oral historian noted: "extraordinary natural events . . . are frequently wrenched from their proper context and connected with local events that seem to make more fitting companions."[35] In this case, the real pillar of cloud and fire was an eruption column from the cinder cone of an Arabian harrat. It erupted—not all at once as Santorini/Thera had—but over the course of weeks or even months, as often happens (in the spring and summer of 2001, Mount Etna in Sicily erupted in precisely this manner). It did not lead the Israelites out of Egypt as claimed in the Bible, but instead was visible only in the final stages of their journey. It is also possible that the erupting magma rose through layers of oil-rich sediments, igniting the oil and causing enormous clouds and fires.[36] As such the eruption column would have been a most impressive sight, smoky by day and fiery by night. Such a feature would have guided, inspired, and intimidated the Israelites as they approached the mountain to renew the covenant with the god of their ancestor, Abraham.

In the course of their journey the Israelites were attacked by the Amelekites who cut off the tail end of their column (Deuteronomy 25:17–18). This encounter became confused with later pitched battles between the two groups (Exodus 17:8–16). Finally, according to Exodus 19:1, the Israelites arrived at the wilderness of Sinai in the third new moon (that is, in the third lunar month) after they left Egypt. In 1628 B.C.E. a new moon fell on January 12 (the first new moon), a second on Feburary 11, and a third on March 12. If the Israelites started their journey in the first few days of February of that year (in the January 12 new moon) their arrival would be after March 12 (in the third new moon). This would be just in time for the first *full* moon after the spring equinox, on March 26.[37]

This arrival, near the end of March, coincided with the time of the wheat harvest in Egypt. The Israelite Exodus at the beginning of the Egyptian barley harvest in early February was the original time for the month of Abib—month of the (freshly ripened) barley (Exodus 13:4: "Today, in the month of Abib, you are going out.")—in the most ancient Israelite calendar, as reflected in the year count quoted by Josephus, mentioned in chapter 1. Centuries later, when the Israelites had transformed themselves into village farmers, the Exodus and the Festival of Unleavened Bread associated with it were shifted to match the barley harvest *in Canaan*. Because of this shift the Exodus became connected with the Israelites' annual covenant renewal sacrifice and accompanying meal (see below and chapter 10), held at the first full moon of spring. The end of the original Exodus journey, at the time of the Egyptian wheat harvest, is reflected in the Festival of Weeks (Shavuot), except that the present festival coincides with the midsummer wheat harvest in Canaan. The tradition that Shavuot marks the anniversary of the giving of the Torah at Sinai thus preserves a vestigial memory of *the original event at Sinai*, at the beginning of spring.[38]

ENCOUNTER AT THE MOUNTAIN OF GOD

When they approached the Mountain of God, the Israelites were met by Moses' father-in-law and his family (probably heading from the coastal plain back to the Hisma), and Moses' father-in-law offered up a sacrifice to Yahweh. When the Israelites finally encamped opposite the mountain, they were told that Yahweh would come down upon the mountain, but that the people should not go up the mountain or many would die: "Be careful not to go up the mountain or to touch the edge of it. Any who touch the mountain shall be put to death. No hand shall touch them, but they shall be stoned or shot with arrows, whether animal or human being, they shall not live" (Exodus 19:12–13). This is a very sensible prohibition to make in the face of an erupting volcano, particularly one that is apparently ejecting pellets or stones (and fits with the Bedouin account of the most recent eruption that was mentioned earlier). The warning "not to touch the edge of [the mountain]" also makes sense if some lava was descending the slope (lava could have been coming from fissures on only one side of the volcano, as is often the case).

On the morning of the third day, there was the sound of thunder, the flash of lightning, the screech of a ram's horn or trumpet, and the mountain was wrapped in smoke and fire: "because the Lord had descended upon it in fire; the smoke went up like the smoke of a kiln, and the whole mountain shook violently" (Exodus 19:18). Shallow, localized earthquakes commonly occur as volcanos erupt. The noise of eruptions, as we have seen, is often compared to thunder, and the screech of steam and other volcanic gases escaping from narrow vents and fissures is not dissimilar to the screech of a ram's horn, which is likewise caused by a gas (air) escaping rapidly from a narrow aperture. Lightning is also characteristic of eruption clouds. Deuteronomy 4:11 describes the eruption this way: "you approached and stood at the foot of the mountain while the mountain was blazing up to the very heavens, shrouded in dark clouds." Psalm 104:32 reflects this tradition: "[the Lord] who looks on the earth and it trembles, who touches the mountains and they smoke."

It has been suggested that these phenomena describe a thunderstorm in high mountains (despite the fact that neither rain nor hail is mentioned) and that lightning striking trees near the timberline produced the smoke and fire.[39] Having personally been through several high-mountain thunderstorms, and one particularly bad one in Montana (which my husband, a native Montanan, described as the worst he'd ever seen), I can say they have nothing in common with the smoke and fire show described in the Pentateuch; nor were they associated with any earthquakes. Rather than being a perfectly normal though violent thunderstorm at some inaccessible mountain altitude, or an elaborate theological metaphor, these accounts in Exodus and Deuteronomy are very specific and very accurate descriptions of a volcanic eruption, as is the pillar of cloud and fire that guided the Israelites to this spot. As one scholar wrote: "it is hard to escape the conclusion that verses like Ex. 19:18 and Dt 4:11 suggest a volcanic eruption," and "No other settled people of the Levant, so far as we know, spoke of divine intervention in these terms. . . ."[40] To people who had never seen or heard such phenomena, they must have been awesome indeed.

REVELATION AT THE MOUNTAIN OF GOD

The biblical narrative of the original revelation on Sinai is a confused account:

> Moses is pictured as ascending and descending Mount Sinai at least three times without any apparent purpose. At times the people are pictured as fearful and standing at a great distance from the mountain, whereas at other times there are repeated warnings which are intended to prevent any of them from breaking forth and desecrating the sacred mountain. . . . God seems to fluctuate between his actually dwelling on the mountain and only descending in periodical visits.[41]

The usual way to explain the difficulties of these texts (particularly Exodus 19–34) is to attribute them to different Documentary Sources (see chapter 4). According to this explanation, the E or northern source deals with the making of the covenant between the people and God, the breaking of the covenant (and the tablets) with the erection of the golden calf or bull, the making of the new tablets, and Moses as a prophet who intercedes with God for the people. The J or southern source is more fragmentary: YHWH appears before the people who have purified themselves; the people stay at a distance because of the danger of being too close to YHWH; Aaron and the elders accompany Moses partway up the mountain, but only Moses goes up to see YHWH.[42] In the Priestly source, Moses goes up to the mountain and God gives him extensive instructions for the tabernacle, the priestly vestments, and other priestly matters. The laws are not given on the mountain but later, when the tabernacle has been constructed. Most scholars agree that this last account (i.e., the Priestly source) is a relatively late version of the Sinai theophany.

At least some scholars agree that the inconsistencies in this account stem from the combining of two ancient traditions while they were still in an oral form: one a tradition of the people seeing God face to face, and the other that of Moses acting as intermediary between God and the people. Y. Avishur sees a series of chiastic parallels in the text, the pivot of which is the Lord descending on the mountain in fire in Exodus 19:18.[43] Chiastic structures, as noted before, are often found in orally inspired works. Many scholars also believe that these passages were originally followed by the presentation of the Decalogue, as described in Deuteronomy 4:10–14:

> how you once stood before the Lord your God at Horeb, when the Lord said to me, "Assemble the people for me, and I will let them hear my words, so that they may learn to fear me as long as they live on the earth, and may teach their children so"; you approached and stood at the foot of the mountain. . . . Then the Lord spoke to you out of the fire. You heard the sound of words but saw no form; there was only a voice. He declared to you his covenant, which he charged you to observe, that is, the ten commandments.

The steep and lofty mountains in the Sinai Peninsula suggested as candidates for the Mountain of God—notably the 2,637-meter (8,455 feet) high Mount St. Catharine and the 2,285-meter (7,467 feet) high Jebel Musa— would have made the comings and goings as described in Exodus 19—24 logistically improbable, especially for the seventy tribal elders. Modern pilgrims scale Jebel Musa only with great difficulty, using a set of steps that were cut into the rock in Byzantine times. One round trip, using stairs that did not exist in Moses' time, takes nearly 4.5 hours.[44] In contrast, many cinder cones of the harrats have relatively gentle slopes only a few hundred feet higher than the land surface on which they sit, although it is unclear whether the cone itself is climbable. Hala'-l-Bedr, for example, is only about 150 meters above a flat sandstone tableland (and 1,500 meters above sea level).[45]

COVENANT SACRIFICE AND MASSEBAH

After his trips up and down the mountain, Moses built an altar at the foot of the Mountain of God and put up twelve standing stones for the twelve tribes of Israel, where burnt offerings of cattle were made to God. Since there weren't twelve distinct tribes at this point of time (see chapter 9), one wonders if twelve might not be an early ritual number among the Israelites. *Massebah*, the Hebrew term for this type of standing stone, are common in the Jordanian and Negev deserts. In ancient times they served as witnesses to treaties and covenants, as markers of sacred areas, or as stones put up to ancestors. The massebah put up by Moses at the Mountain of God had elements of all three of these uses. Usually massebah come in groups of two or three, occasionally in groups of five, seven, or nine.[46] At the foot of Hala'-l-Bedr, according to Musil's informants, there were twelve, the same number said to have been erected at the foot of the Mountain of God (Exodus 24:4).

After examining photographs taken in and around Hala'-l-Bedr, Professor Jean Kœnig claimed he had found the massebah, a pile of sandstone rocks near the foot of what he believed to be Hala'-l-Bedr.[47] His conclusions were subjected to a withering critique by Jacqueline Pirenne, who claimed that Kœnig's photographer had missed Hala'-l-Bedr entirely and that the rocks mentioned by Musil were the same as the red granite ruins observed by the nineteenth century explorer Charles Doughty, not the sandstone formation noted by Kœnig.[48] Granite rocks are exposed no closer than about twenty-four kilometers from Hala'-l-Bedr, and "ruins" is a term that usually implies destroyed structures of some kind rather than free-standing stones; thus it seems unlikely that Doughty's red granite ruins are Musil's massebah.[49] Whether the photographer, and thus Kœnig, examined the right volcanic cones (Hala'-l-Bedr and al-'Asi, the latter of which seems to have acquired a new name) and the right pile of sandstone blocks is certainly doubtful.[50] If this is not the case, Musil's massebah have not been identified and described. Even if they still exist, wind scour and sand blasting would probably have removed any traces of human alteration.[51]

Many people have been reluctant to consider a location for the Mountain of God in Arabia because its most common name, Mount Sinai, seems to connect it with the Sinai Peninsula. The name Sinai, in one hypothesis, was derived from the Sumerian/Akkadian moon god Sin, who gave his name to the Wilderness of Sin, one of the earlier stops in the wilderness wanderings of the Israelites, and to Sinai, the peninsula, and to Sinai, the holy mountain.[52] Hala'-l-Bedr means, in English, "crater of the full moon,"[53] but because it is in Arabia, and not in the Sinai (or near the Wilderness of Sin for that matter), these places do not seem to have a common name connection.

Another hypothesis derives the name Sinai from the Hebrew word for "bush," *sêneh*. In other Semitic languages the cognate word refers to a particular thorny shrub, and a species of the thorny acacia is the best candidate for the biblical bush.[54] *Acacia tortilis*, known to the Arabs as "samr," marks the two-to-four-inch (five-to-ten-centimeter) rainfall zone. Because it does not grow with less than two inches of rainfall annually, it is an important moisture indicator in the desert. No other bush remotely resembles it, and it grows extensively and often exclusively in the area of the northern Arabian harrats and throughout the region.[55] In moister areas (with at least four inches of rainfall annually), other species of acacia will also grow. In these wetter areas the acacia is recognized as a tree and is known by another word. In the seventeenth century B.C.E., the Israelites would have used the word sêneh to refer to *Acacia tortilis* in its bush form, but since in many dry regions it would have been the only type of bush growing, sêneh would have meant "bush." This would explain why widely separated desert areas were named sêneh: the Sinai Peninsula and the Wilderness of Sin. On the sandstone tableland of Hala'-l-Bedr (that is, above the wadi bottoms and valley floors such as the al-Ǧaw basin), rain was the only source of water, and only hardy vegetation such as the sêneh bush would have been able to grow.

Naming the Mountain of God for a bush, especially one that marked Moses' first encounter with God, is just the sort of thing people of this region would do. The Hebrew scriptures contain a number of places named for streams, trees, and the like. The Medieval Arab geographer Yakut wrote, "It is said that *Sina'* is the name of its [the mountain of God's] rocks or its trees," and "But in the Nabataean language every mountain is called Tur and as soon as bushes and trees grow on it, it is named 'Tur

Sina'.'"[56] Musil found a number of geographic features (usually valleys or watercourses, and one hill east of the Wadi el-Arabah) named for another kind of bush, a type of gorse, the *Ratam*, *Retama*, or *Retame*: "a shrub with long rather stiff branches, long needle-shaped leaves, and hanging scented flowers."[57] At least two wadis originating in the Harrat ar Raha are named Retame. Its Hebrew equivalent is *ritmâh* or *rithmah*. In Numbers 33:18, Rithmah is a stop on the wilderness journey of the Israelites, only three resting places away from "the wilderness of Sinai" (Numbers 33:15, 16).

There is a second name given to the Mountain of God in a few passages—Horeb. Horeb has the general meaning of a desert region (*hrb*) and may have originally meant only the desert region in which the mountain was placed.[58] But hrb would also be the written form of Mount Harb, one of two peaks (the other is Dibbah) that are notable landmarks in the area and mark the place where Israelites would have turned off the caravan route to cross into the volcanic desert of Harrat ar Raha when journeying to Hala'-l-Bedr.[59] If, as Noth believes, Horeb is a late addition to an older tradition, Horeb may simply be a form of Harb, the name being transferred from the landmark mountain to the Mountain of God.

PILGRIMAGES TO THE MOUNTAIN OF GOD

Numbers 33:5–49 presents a list of stops supposedly followed by the Israelites on their journey from Egypt to the plains of Moab. This itinerary was thoroughly studied by Graham Davies, who concluded that it formed the basis for the other itinerary segments in Exodus and in Numbers 20 and 21, and that it probably described an actual, widely known route.[60] Earlier, Martin Noth suggested that part of the Numbers 33 itinerary was a pilgrimage route to the Mountain of God.[61] Both Noth and Jean Kœnig discovered that the names in Numbers 33 near "the wilderness of Sinai" correspond to names (and in the same order) in or near Harrat ar Raha. Using these names and Musil's map, Kœnig even traced a route from Mount Harb (which he equated to Mount Shepher in Numbers 33:23) to Hala'-l-Bedr that follows an ancient track.[62]

In Deuteronomy 1:2 it is stated that it is eleven days' journey from Horeb to Kadesh (for the location of Kadesh, see chapter 6). Davies notes that the standard ancient rate of travel is about thirty kilometers per day, a distance that would better fit the distance between Kadesh and Mount Harb than that between Kadesh and Hala'-l-Bedr.[63] However, Alois Musil's examination of Arabic sources detailing the pilgrim routes to the holy cities of Mecca and Medina, as well as the study done by Greta Hort, show an average travel distance of fifty-five to seventy kilometers (an average of thirty *miles*) per day through the Hegaz. The distance between Kadesh and Hala'-l-Bedr would fall within this thirty mile per day range.[64] This latter figure involves traveling on camels, which did not become domesticated until late in the second millennium B.C.E.; thus the statement in Deuteronomy 1:2 probably reflects a later time, such as Elijah's, when travel through the desert was typically done on camels and when the travelers were pilgrims, not the group led by Moses who, with their herds and their elderly and children, would have traveled at a much slower pace.

It is clear that the Mountain of God, where the first divine revelation was received, remained a sacred and holy place to the Israelites, one they would return to again and again throughout their sojourn in the wilderness (see chapter 6) and in later centuries, as pilgrims. Early on, the mountain was the focal point of the Israelites' religious and tribal identity. However, as the years passed, Moses died, the volcanic emissions ceased, and the Israelites moved north, the sacredness of the mountain apparently was transferred at least in part to the tabernacle and the ark within it (see chapter 6). This transfer paved the way for the establishment of cult centers in Canaan far distant from the Mountain of God. In time, the Mountain at Sinai lost its importance as a pilgrimage destination for the Israelites and became only a sacred memory.

CHAPTER SIX

The Sojourn in the Wilderness

A LAND FLOWING WITH MILK AND HONEY

After their first stay at the Mountain of God, Moses led the Israelites north, intending to settle in Canaan. He sent spies ahead to scout out the land. The spies reported back to Moses at Kadesh that "We came to the land to which you sent us; it flows with milk and honey, and this is its fruit. Yet the people who live in the land are strong, and the towns are fortified and very large" (Numbers 13:27–28).

Southern Canaan at this time had a number of cities and towns, both large and small. The most impressive features of the larger Canaanite cities were their massive fortifications. Surrounding walls were abutted by enormous ramparts, and access to the city was only by complex gates that provided effective defense.[1]

The largest sites were closest to the coast, often deliberately built at the mouths of rivers or wadis to take advantage of the flourishing maritime trade between the Nile Delta, the northern Levantine coast, and Cyprus and the Aegean.[2] The largest was probably Tell el-'Ajjul, thought to be the Sharuhen (see figures 5.4 and 9.1) mentioned in the Egyptian texts. Manufactured goods from Egypt, ceramic wares from Cyprus, and exotic metals such as gold and tin were brought to the large coastal towns by ship and transferred to smaller towns inland. These inland towns were surrounded by villages and hamlets that produced wine, olive oil, and other agricultural products for trade. Southern Canaan was a major barley-growing area, and also provided honey and herbs as well as grapes and olives.[3] There was probably a trade in cattle and products from the herds of sheep and goats.

The larger towns were also manufacturing centers, for numbers of locally made gold and ivory objects appear to have been fashioned on the spot. The gold and ivory came originally from East Africa, and was passed either up the Nile to Avaris or overland by donkey caravan. The ancient overland spice route from southwestern Arabia may also have had its northern terminus in the cities and towns of southern Canaan, especially after the Wadi Tumulat was abandoned.[4] Aromatics, especially myrrh from Punt (East Africa and Ethiopia), may have also come up this overland route after crossing the narrow stretch of sea to Yemen, thus avoiding the endemic piracy in the Red Sea.[5]

The Israelite spies also reported to Moses: "The Amalekites live in the land of the Negeb; the Hittites, the Jebusites, and the Amorites live in the hill country; and the Canaanites live by the sea, and along the Jordan" (Numbers 13:27–29). The Israelites had already fought the Amalekites, one of the pastoral peoples living in the area of the Negev. The term Hittite in the Bible seems to be more of a geographic term than an ethnic one, generally referring to groups of northern people and specifically to the Hurrians, a non-Semitic people who migrated south into Syria and Canaan in the seventeenth century B.C.E.[6] The term Jebusites—in this context, at least—may also refer to a group of Hurrians, since Aranuah the Jebusite who sold the threshing floor to King David bore a Hurrian-style title.[7] A clay tablet from Tell Rumeida (Hebron) reveals both Amorite and Hurrian names.[8] The Amorites are well-attested migrants into Canaan in the earlier part of the Middle Bronze Age,[9] while the Canaanites were the original, pre-MB inhabitants of the area, still occupying the coastal plains and the Jordan Valley. From these pieces of information we can see that the spies' description preserves a surprisingly accurate picture of Canaan during the latter part of the Middle Bronze Age.

At Hebron (Tell Rumeida) the spies reported that Ahiman, Sheshai (Sheshi), and Talmai, the Anakites, were there (Numbers 13:22). Sheshi, as mentioned in chapters 2 and 5, was the first Hyksos ruler, whose scarabs were found along the Nile as far south as Kerma in Nubia and in several southern Canaanite cities, including Jericho. One Sheshi scarab was found in a tomb at Tell el-Maskhuta.[10] Clearly, the Israelite oral tradition preserved his name and a broadly correct time period.

ATTEMPTED INVASION

Archaeologists have found numbers of bronze weapons, particularly battle axes, within these Middle Bronze Age Canaanite cities. There is little doubt that their rulers could have marshaled large and effective fighting forces from both their towns and the surrounding countryside (over which they exercised effective political control) to defend themselves against nomadic invaders.[11]

Based on estimates of the nineteenth century Bedouin population of the Wadi Tumulat, I have suggested a figure of three thousand Israelites, supplemented by possibly as many as two hundred Avarans. The biblical number of 603,550 fighting men[12] is a typical oral historical exaggeration. Given approximately four to five noncombatants for each fighting man, this number would yield a total of over three million Israelites, more than the entire population of Egypt at the time! However, *six hundred* combatants, if related to the same number of noncombatants, is not wildly different from the number who fled the Wadi Tumulat. Six hundred combatants would have had no chance against the armies of the Canaanite city-states and their Amalekite allies. Numbers 14:44–45 describes the Israelites' abortive invasion into the southern Judean hill country, and how they were thrown back. They retreated into the desert and began their extended sojourn in the wilderness.

EARLY DISSENT

Although the biblical version states that the original attempt of the Israelites to invade Canaan from the south failed because they were going against God's will, this was a later theologizing to explain why the invasion failed. The predictable result of this failure was dissent and a splintering off from the main group.

In Numbers 16 the Reubenites Dathan, Abiram, and On and their families separate from the main body of the Israelites and say to Moses: "We will not come! Is it too little that you have brought us up out of a land flowing with milk and honey [here they mean Egypt] to kill us in the wilderness, that you must also lord it over us? It is clear that you have not brought us into a land flowing with milk and honey [here they mean Canaan]. . . . We will not come!" (Numbers 16:12–14). Another revolt, led by the Levite Korah, supposedly happened at the same time but, though Korah was said to have been swallowed up by the earth along with the Reubenites, all his sons survived, an indication that the Levitical uprising took place at a different time (see table 6.1); later, the two revolts were fused together.[13] Greta Hort has suggested that the dissenting Reubenite splinter group was camped on a *kewir* in the southern part of the Wadi al-'Arabah (see figure 5.4). A kewir is a mudflat with thin layers of mud

TABLE 6.1.
Revolts in the Wilderness

Reubenite revolt	The families of Dathan, Abiram, and On leave the main body of Israelites and are swallowed up by the earth.	Numbers 16:1*b*, 12–14, 15, 23–32*a*, 33–34.
Revolt of the Levites under Korah, Moses' cousin	Korah and 250 Levites or Israelites protest the elevation of Aaron, asserting that all the congregation is holy. The protestors are told to bring censers to the tent of meeting, where they are struck down by fire that comes out from the Lord. Variant in Leviticus (incorrectly) identifies Aaron's sons Nadab and Abihu as the censor-carrying would-be priests who are struck down by fire from the Lord.	Numbers 16:1*a*, 2–7, 11*b*, 16–21, 35, 40 (see also Numbers 16:41–50). Leviticus 9:23–24; 10:1–2.
Miriam and Aaron speak out against Moses	Miriam and Aaron speak out against Moses because of the Cushite woman Moses had married. They also claim prophetic authority along with Moses. Miriam becomes leprous and is banished from the camp for seven days.	Numbers 12:1–2, 4–16.
Aaron's rebellion	At Rephidim, the Israelites quarrel with Moses because they have no water. Moses goes on ahead to the Mountain of God with some of the elders and strikes a rock, and water gushes forth. Returning to the Israelite camp, he finds that Aaron has taken the people's gold and made it into a golden bull, built an altar to it, and made sacrifices to it as the people danced around. The first set of tablets from God are broken. Moses calls the Levites to his side and they kill those who were worshiping the golden bull. The bodies of Aaron's sons Nadab and Abihu are dragged by their tunics from the camp by their kinsmen. Moses takes the golden bull and grinds it up and scatters it on the water or stream. Aaron is stripped of his priestly vestments and executed at the Mountain of God. Aaron's son Eleazar becomes priest.	Exodus 17:1–7 (see also Numbers 20:2–11, 13). Exodus 32:2–6; Deuteronomy 9:16. Exodus 32:19*b*; Deuteronomy 9:17. Exodus 32:19*a*, 25–29; Deuteronomy 33:8. Leviticus 10:4–5. Exodus 32:20; Deuteronomy 9:21. Numbers 20:24–28.

and salt above an expanse of soft clay. When pressure, such as tents, tent pegs, people, and animals, was put on the surface it would break up and collapse.[14] Another possibility is an earthquake, a relatively common occurrence in the tectonically active Wadi al-'Arabah.

THE LENGTH OF THE SOJOURN

After the seemingly miraculous destruction of the Reubenite dissenters, Moses' authority was restored and the Israelites began their extended sojourn in the wilderness. According to the scriptures this sojourn lasted forty years. But the interval between a 1628 B.C.E. Exodus date and the destruction of Jericho in about 1550 B.C.E. (see chapter 8) is seventy-eight years, not forty. In both the Hebrew and Christian scriptures the number forty is applied to all sorts of unknown time intervals. Forty is a "perfect" or a religious number.[15]

Moses was said to have been eighty years old when he confronted Pharaoh (Exodus 7:7) and traditionally to have been about forty when he fled to Midian. He then lived for forty years in the wilderness, dying at age 120 (Deuteronomy 31:2). More realistically, an impulsive killing such as Moses carried out in Egypt is the sort of thing a young man would do, say one in his early twenties. The highly distorted story in Exodus 4:25, in which Moses' wife Zipporah cuts off her son's foreskin to save Moses from being killed by God, implies that Moses' son was in early adolescence when Moses set out again for Egypt, for circumcision is often an adolescent rite of passage as it was for the ancient Egyptians.[16] If so, Moses would have been in his mid- to late thirties when he returned to the Delta and led the Israelites out of the Wadi Tumilat. Psalm 90:10—this psalm is called a prayer of Moses—says "the days of our life are seventy years or perhaps eighty if we are strong," implying that Moses lived well into his seventies. The thirty-eight years the Israelites spent traveling from Kadesh-barnea (Petra) to the Wadi Zered (the Wadi el-Hesa—see figure 5.4) (Deuteronomy 2:14) may actually reflect the years *Moses* spent in the wilderness before he died, and the sacred number forty may have originally been the numbers of years the Israelites spent in the wilderness *after Moses' death*.

The scriptures do retain some hints of the actual, longer sojourn in the wilderness. In Numbers 14:29 and 32 God says that "your dead bodies shall fall in this (very) wilderness," a reflection of the fact that everyone alive at the time of the Exodus (not just those over twenty, a later rationalization) did die in the wilderness. A second indication is the frequent statement that God will punish children for the iniquity of their parents to the third and fourth generation.[17]

The most direct indication of the true length of the sojourn is found in Genesis 15:13–14, 16, a supposedly prophetic but obviously later description of the Exodus and sojourn, which describes how Abraham's descendants will be slaves in an alien land for four hundred years and, after leaving Egypt, shall come back to Canaan in the fourth generation. Allowing about twenty-three to twenty-five years per generation brings us to sixty-nine to seventy-five years for the first three generations, so that a circa seventy-eight-year sojourn fits comfortably in the fourth generation. There is also a genealogical hint in the family of Caleb, one of the two faithful spies during the first, unsuccessful penetration into Canaan. Caleb's great-grandson, Bezaleel son of Uri son of Hur, makes the ark to hold the tablets of the Decalogue, another four-generation span.

The longer, actual length of the sojourn in the wilderness meant that Moses died before the Israelites were able to return to Canaan. This created a need in later times to explain why Moses, God's chosen leader, did not get to lead the Israelites into the promised land. One explanation was that Moses was being punished because both he and Aaron had disobeyed God at the waters of Meribah. A second, quite different reason was offered in Deuteronomy 1:37: God would not allow Moses into the promised land because Moses had not trusted enough before the first attempt to penetrate Canaan and instead sent spies to scout out the territory. Both explanations are nonsensical.

Unfortunately, telescoping the longer timespan of three-plus generations into the shorter perfect number of forty eliminates one of the story's truly remarkable features—how the Israelites remained committed to moving back to Canaan despite the death of their great leader, Moses. Fortunately, there are a few indications in the scriptures themselves of how this commitment was maintained.

ANNUAL COVENANT RENEWAL GATHERINGS AT THE MOUNTAIN OF GOD

Certain scholars have suggested that the oldest story of the Sinai covenant is found among the verses in Exodus 24, 34, and 32.[18] In fact, these passages are a collection of later episodes that were fused together with the first visit of Moses and the Israelites to the Mountain of God.

In the first episode, Moses, Aaron, Nadab, and Abihu and the seventy elders went up to the mountain and had a covenant meal, reminiscent of the covenant sacrifice and its accompanying meal between Jacob and Laban in Genesis 31:44–54. It is likely that this meal on the mountain reflected a regular sacrifice and covenant meal, made by the Israelites each year of their sojourn in the wilderness. As Siegfried Herrmann noted: "Possibly in Exodus 24:9–11 we have the rudiments of an early sacral tradition which has been preserved quite by chance; it would seem to know of sacred events and a solemn festival at the mountain, but as a periodic custom rather than as a single happening."[19] Numbers 9:1–5 and the first lines of Psalm 81 also suggest that the Israelites gathered at the Mountain of God to offer up their covenant renewal sacrifices at the first full moon of the spring. Here the foundation story of the Exodus was repeated and renewed, each year, keeping it fresh in the minds of the descendants of those who had experienced God's "signs and wonders."

Exodus 24:9–11 also describes a partial transfer of leadership from an aging Moses to a younger Aaron, a good many years after the first visit to the Mountain of God. The mountain, where Moses went up every spring at the covenant renewal gathering, meditated, prayed, and listened for God, would have been the obvious place to transfer any divine authority. In Exodus 4:14 Aaron is simply the Levite brother of Moses, i.e., a fellow Levite.[20] Only later does Aaron become Moses' brother and a full participant in the Exodus story, equal to Moses, a change that mirrors the way the role of the tribal leader in the Hopi story (mentioned in the Introduction) expanded through time.[21]

It is noteworthy that there is no mention of fire or smoke in these verses, but only something like sapphire tiles beneath God's feet. The volcano, then, was not erupting into the air at this later time. The basalts of the Harrat ar-Raha are unusually rich in <u>olivine</u> crystals (<u>phenocrysts</u>), which nineteenth century explorer Charles Doughty termed "common greenish volcanic crystals."[22] I myself have seen blue olivine crystals from a Hawaiian lava core. Optical refractive characteristics of the olivine are responsible for this unusual color.

REVOLT OF THE LEVITES

Many of the Levites, led by Korah, Moses' cousin, opposed the appointment of Aaron (see table 6.1). Hearing the protest, Moses directed Korah and 250 dissidents to appear the next morning at the tent of meeting, each one bringing with him a censer filled with fire with incense on it. Aaron too appeared with a censer, but Moses and Aaron were directed by God to move away from the rebellious Levites. Once they had done so "fire came out from the Lord" and consumed the 250 men. A variant version of this story, in Leviticus 10:1–7, has Aaron's sons Nadab and Abihu taking their censers and offering unholy fire to God, who then consumed them with fire. Greta Hort has suggested that the protesting Levites were struck down by lightning attracted to the metal in the censers when a desert thunderstorm came up.[23] However, the fire "came *out* from the Lord," it did not fall from heaven as it did in a similar incident when fire consumed the offering of Elijah on Mount Carmel in his contest with the priests of Baal.[24] Molten lava from alkaline basalt volcanoes can break out through surface vents and fissures, sometimes miles distant from the crater itself, and flow across the ground at will. This may have been the fire that came out from the Lord. In addition, along with the lava there could have been deadly gases issuing from these same fissures, with similarly lethal effects to those gathered nearby.

THE SHARING OF LEADERSHIP BETWEEN AARON AND HUR

At another, later, covenant renewal gathering at the Mountain of God, Aaron now shares leadership duties with Hur, for Moses says to the elders: "Aaron and Hur are with you; whoever has a dispute may go to them" (Exodus 24:14*b*). Hur, the son of Caleb, is the equal of Aaron here and in Exodus 17:10–12, where he and Aaron hold Moses' arms up during the battle with the Amalekites.

Caleb, Hur's father, is given two genealogies: he is either a son of Hezron, who is a grandson of Judah, or he is the son of Jephunneh the Kenizzite.[25] In Judges 1:13 and 3:9 Kenaz is Caleb's younger brother, but in Genesis 36 Kenaz is a son or clan of Esau. Many clans or peoples living south of Cannan are connected to Esau in the biblical genealogies, and in this case Kenaz or Kenizzite is probably the equivalent of Kenite. The Kenites were smiths or metal workers living in the Wadi al-'Arabah[26] and would have had the skill to construct ritual objects of bronze or gold. Bezaleel, Hur's grandson, makes or oversees the making of the ark and all the furnishings of the tent, the incense and burnt offering altars, the table, vessels, and vestments. Emphasis is on the gold overlay or the fashioning of these objects from gold.[27] Elsewhere, however, the altar Bezaleel constructed is (more realistically) made of bronze, not gold.

In Numbers 10:29 Moses asks his brother-in-law, Hobab the son of Reuel the Midianite, to accompany the Israelites: "Do not leave us, I pray

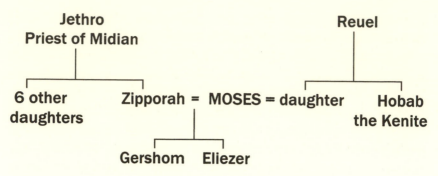

FIGURE 6.1. Proposed reconstruction of Moses' relationships to individuals mentioned in Exodus 18:1–12, Numbers 10:29, Judges 1:16–17, and Judges 4:11.

you, for you know how we are to encamp in the wilderness and you will serve as eyes for us." Judges 1:16–17 says the descendants of Hobab the Kenite went up with the people of Judah from the city of palms ('Ain Hosb, in the Wadi al-'Arabah not far from Petra) into the Negev near Arad. It would seem that one tradition remembered Hobab as Moses' brother-in-law and thus a Midianite and another, probably more accurately, remembered Hobab as a Kenite.

I believe these narrative remnants about Caleb and Hur and their family, and about Hobab, point toward an alliance between the Israelites and the Kenites. The memory of this alliance is preserved in 1 Samuel 15:6 when King Saul says to the Kenites: "for you showed kindness to all the people of Israel when they came up out of Egypt." This alliance was sealed by Moses taking a daughter of Reuel the Kenite as a second wife, his first wife having been a daughter of the Midianite priest Jethro (see figure 6.1). Hur's place as one of Moses' assistants was evidently part of this alliance, which included military actions against the Amalakites.[28] In Numbers 31:8 Hur is one of the five leaders of Midian killed by the Israelites, a possible indication that Hur died of the plague outside Jericho (see chapter 7). For nationalistic reasons, later Israelite tradition made the family of Caleb, Hur, and Bezaleel into members of the tribe of Judah and gave them Judahite genealogies (see chapter 9).

AARON'S REBELLION

At some point Aaron, with the help of his sister, Miriam the prophetess, attempted to take control from Moses. Aaron and Miriam justified their criticism of Moses by asking, "Has the Lord spoken only through Moses? Has he not spoken through us also?" (Numbers 12:2). Their attempt to secure control was easily put down by Moses, and the punishment was merely short-term banishment (probably for both of them, not simply Miriam). But a later incident, recounted in Exodus 32, was a full-scale rebellion.

In Exodus 32 while Moses is on the Mountain of God, Aaron takes gold from the Israelites and makes it into the image of a young bull (not a calf) and the people worship it. When Moses finally comes down from the mountain with the newly made tablets of the covenant he finds the people running wild—or rioting. Moses breaks the tablets of the covenant, destroys the statue of the young bull, and makes the people drink the ground remains of the idol. After interrogating Aaron, Moses calls the Levites to his side and they go among the Israelites and kill a good many of them. The next day God sends a plague on the rest of the people.

This account has inconsistencies and duplications which have led some scholars to presume that it is has been composed from disparate, often late, sources, but recent analyses by Ralph E. Hendrix and Christine E. Hayes have pointed out the chiastic structure of the story (and noted the frequent use of repetition, both characteristics of orality) as well as the internal unity of the whole passage. Rather than being derivative, Exodus 32 is the source for other biblical passages that refer to this incident.[29]

By any logic, Aaron should have been punished for this idolatry for "the Lord was so angry with Aaron that he was ready to destroy him" (Deuteronomy 9:20), but only in Numbers 20:2–13 is Aaron punished, because he assisted Moses when Moses disobeyed God at Meribah by striking the rock to get water instead of *speaking* to it! Meribah here is supposed to be at Kadesh.

The story of Meribah in Numbers 20:2–13 is really the first part of the story of the death of Aaron in Numbers 20:24–29. It comes directly after the story of Miriam's death at Kadesh and was attached to that story at a very early time, when storytellers were relating the deaths of first one and then the other sibling: "and Miriam died at Kadesh, . . . and Aaron died."[30] Thus the incident of the quarrel at Meribah, alluded to rather mysteriously in a number of later Biblical passages,[31] was juxtaposed with Kadesh early in oral tradition because of the association of Miriam and her death (at Kadesh) with her brother Aaron and his death.

The same incident is related in Exodus 17:1–7, but in this passage Meribah is at Rephidim. According to this text, the people were camped at Rephidim, which according to Numbers 33 is only one stopping place from the Mountain of God (see figure 5.2, inset). They complained to Moses because there was no water, and Moses asked God, "What shall I do with these people? They are almost ready to stone me." God instructed Moses to go on ahead with some of the elders of Israel and God would be standing in front of them on the rock of Horeb. God directed Moses to strike the rock, and water would come out of it, so that the people could drink. Moses and the elders went on to the Mountain of God as directed, Moses struck the rock in the presence of the elders—and the story ends abruptly with an explanation for the names Massah (test) and Meribah (quarrel).[32]

What actually happened next is described in Exodus 32:25–29: Moses returned to the Israelite camp and found the full-scale rebellion fomented by Aaron: "For Aaron let them [the people] run wild, to the derision of their enemies." The chiastic structure of Exodus 32 reveals that the passage (verse 26a) where Moses stands at the gate of the camp and calls on all who are on the Lord's side to come to him is the focal point of the whole story.[33] In verse 26b the sons of Levi come to his aid. Many Levites, remember, had never been happy with Aaron's appointment in the first place. Deuteronomy 33:8, where Moses blesses the tribe of Levi, carries a remnant of this original version, for there Moses says: "Give to Levi your Thummim and your Urim to your loyal one, whom you tested at Massah, with whom you contended at the waters of Meribah." In short, the Levites had stayed loyal to Moses—and thus to God—at Meribah, just as described in Exodus 32:26 when they put down Aaron's rebellion.

Two of Aaron's sons, Nadab and Abihu, were killed in the rebellion, and Moses called on their kinsmen to drag their bodies *by their tunics* from the camp (see Leviticus 10:4–5; in these verses Nadab and Abihu were presumably burned to death but still had their tunics intact!). But what became of Aaron? In Numbers 20, the Israelites set out from Kadesh (actually, Rephidam) to Mount Hor (actually, the Mountain of God) and God said (Numbers 20:24): "let Aaron be gathered to his people. For he shall not enter the land that I have given to the Israelites, because you rebelled against my command at the waters of Meribah." Moses, Aaron, and his son Eleazar "went up Mount Hor [actually, the Mountain of God], in the sight of the whole congregation. Moses stripped Aaron of his vestments and put them on his [Aaron's] son Eleazar [who obviously had not taken part in the rebellion]. And Aaron died there on the top of the mountain" (Numbers 20:27, 28)—that is, Aaron was executed at the Mountain of God for his rebellion.[34] Because of erroneously placing Meribah at Kadesh, and confusing Mount Hor with Mount Horeb, the Mountain of

God, and further conflating the rebellion of Korah with, first, the Reubenite revolt and, second, the role Nadab and Abihu played in their father's rebellion, these stories became fragmented and their original meanings lost (see table 6.1).

THE TABLETS OF THE TEN COMMANDMENTS

Deuteronomy 4:13 makes it clear that the tablets with the Ten Commandments written on them were made directly after the first appearance of God to the Israelites at Mount Sinai. The tablets would have been carried with the Israelites as they wandered in the wilderness. Returning to the mountain each spring, the tablets would have played a key role in the covenant renewal rites. Thus, even though Moses supposedly broke the first set of tablets when he saw the golden bull and the dancing,[35] the tablets almost certainly were broken by Aaron's supporters during the rebellion. Monumental stone inscriptions in the ancient Near East served to "eternalize an event," even to people who couldn't read the words. Egyptians in the Middle Kingdom (1900–1750 B.C.E.) created images of their enemies in stone, terra cotta, or wood, or wrote their enemies' names on pottery. The image or name was then cursed and broken.[36] It is scarcely believable that Moses would break a tablet with God's name on it, but the breaking of the tablets by the rebels would destroy the concrete symbol of the covenant and of Moses' authority from God. After Aaron's rebellion was put down, a second set of tablets was made, and the Levites, who had proved their loyalty, were given the task of guarding the tabernacle or tent that held the ark with the second set of tablets and ordered to kill anyone who came near it. The guarding of the ark suggests that some of the people were seen as potential threats to the new tablets.

In Deuteronomy 10:1–5 Moses says he himself made the ark immediately upon coming down with the second set of tablets, but in Exodus 37:1 Bezaleel made the ark at a slightly later time. If, as I maintain, there was an interval of years between the making of the first and the second sets of tablets, then these two stories of the making of the ark may actually refer to two arks, the first made by Moses for the first set broken during Aaron's rebellion, and the second ark made later by Bezaleel for the second set of tablets, guarded by the Levites.

The ten words of the covenant would have been written on slabs of the soft sandstone that is found throughout the region. In future centuries many different peoples of the Hejaz would use this same sandstone for thousands of their own inscriptions. As nineteenth century English explorer Doughty noted: "We see in the cliff-inscriptions at Medain, that the thickness of your nail is not wasted from a face of soft sandstone, under this climate, in nearly two-thousand years!"[37] We know that both the Fourteenth Dynasty people and the Hyksos used Egyptian hieroglyphics, and that proto-Canaanite script existed as well. Moses, or someone in the group, could have had the knowledge to inscribe words on stone.

THE KEEPING OF THE COVENANT AND THE RETURN TO CANAAN

The third and latest episode at the Mountain of God described in Exodus 24, 34, and 32 comes from even later in the wilderness sojourn, probably not long before the death of Moses. Moses and his young servant Joshua go up alone to the Mountain of God (Exodus 24:13a), and Joshua promises to keep the covenant to return to Canaan and not make covenants with the Canaanites.

Joshua son of Nun is named as one of the original spies Moses sent to reconnoiter Canaan prior to the original attempt to penetrate the country, but this is chronologically impossible. Numbers 13:8 gives the name of the original Ephramite spy: Hoshea son of Nun. Numbers 13:16 says "Moses changed the name of Hoshea son of Nun to Joshua," the story's way to explain the conflation of the two individuals. Joshua, if he was actually Moses' young servant, seems later to have been the Israelites' war-leader against the Amalekites. The long, protracted battle against the Amalekites described in Exodus 17:8–13 is probably a folkloric version of a whole series of battles and skirmishes between the Israelites and the Amalekites during their sojourn in the wilderness, some encounters going one way, some another. Joshua would thus have been middle-aged by the time the Israelites under his leadership made another attempt to enter Canaan. This time, with the possible exception of some of the groups that eventually became part of the southern tribe of Judah (see chapter 9), the Israelites did not attempt a southern penetration but instead moved north on the eastern side of the Wadi Arabah, outside Amalekite territory, around the eastern side of the Dead Sea, and through what later became the kingdoms of Edom and Moab.

THE ISRAELITE JOURNEY THROUGH EDOM AND MOAB

One of the most compelling arguments against an early date for the Exodus is the dearth of any substantial settlements in Edom and Moab that correspond to the biblical accounts of the Israelites' journey through these areas (see figure 5.4). In the 1930s and 1940s, an American, Nelson Glueck, conducted a site survey east of the Dead Sea and the Jordan River in what is today's kingdom of Jordan and found hardly any archaeological remains that could be dated to the Middle or Late Bronze Age. Over fifty years later these findings are still valid: there is "scant evidence" of Middle and Late Bronze Age settlements in Moab and an "occupational gap" in MB-LB Edom.[38]

This does not mean that these areas were unoccupied at that time, however. It simply means that most of the people who occupied these territories were archaeologically "invisible," as Israeli archaeologist Israel Finkelstein has demonstrated.[39] Nomads living in tents and using vessels made of skins and utensils of wood would leave virtually no archaeological remains. Early nomadic leaders were not kings but rather tribal chieftains. Eventually some of these tribes would develop into the tribal kingdoms found in the stories in the Hebrew Bible.[40] Stories of the Israelites' later conflicts with these tribes and their kings would eventually be included with the original Israelite journey north. This is an example of a common form of anachronism that oral historians call the "lightning rod effect," where later events accumulate around an earlier "time of origins."[41]

Egyptian texts from the nineteenth century B.C.E. mention a land called Shutu associated with "the sons of Sheth," in the region that later became Moab; in Numbers 24:17 the term Shethite is synonymous with Moabite. Even earlier, the Egyptian story of Sinuhe (c. 1900 B.C.E.) mentions a mountain chieftain of Kushu named Ya'ush. Kushu was south of Shutu, in the mountainous area (later also called Seir) that later became known as the homeland of the Edomites, who are equated in the Bible with the sons of Esau. Remarkably, the Ya'ush in Sinuhe's tale is found in Genesis 36:5 and 18 as Jeush (Ye'ush), a son of Esau.[42]

The term Kushu or Kushite is also found in the story of Moses, for in Numbers 12:1 Aaron and Miriam speak against Moses for marrying a Cushite (that is, Kushite) woman, although elsewhere Moses' wife is described as a Midianite. In Habakkuk 3:7 the tents of Cushan are equated with (or used in a parallel sense to) the tent-curtains of Midian. Obviously, at some point, the Israelite tradition equated Midian with Kushu, although Kushu or Cushan was inhabited by descendants of Esau, who were later equated to the Edomites. The fusion of Midian with Kushu (later the land of Edom), and of Edom with Seir may be why these lines appear in the ancient Song of Deborah: "Lord, when you went out from

Seir, when you marched from the region of Edom . . . The mountains quaked before the Lord, the One of Sinai. . . ." (Judges 5:4–5).

In the earliest version of the Israelite journey, in Numbers 33, the Israelites proceed north from the Gulf of Aqaba (Ezion-geber), through the territory of the sons of Esau (it is called Seir, not yet Edom) to Moab. This list of stopping places is just the sort of abbreviated version one would expect to find in oral tradition carried down for centuries.

On significant stop was at Kadesh. The third century c.e. Christian scholar Eusebius in his *Onomasticon* describes "Kadesh-Barnea" as "a wilderness that stretches at the town of Petra in Arabia; there Miriam ascended and died (Numbers 20:1)."[43] Josephus in the first century places Miriam's death at Mount Zin. The word Zin, which refers to something sharp, probably applies to jagged mountain peaks, similar to those found in the area of Petra and west of the Wadi al-'Arabah in general.[44] From Kadesh the Israelites head north across the Wadi Zered (Wadi el-Hesa) and go through the territory of the descendants of Lot, known as Ar or Moab.

In Moab the Israelites stopped at Iye-abarim (possibly 'Ay, ten kilometers southwest of Kerak) and at Dibon-gad or Dhiban, north of the Wadi el-Mūjub (see figures 5.4 and 9.2). Although no archaeological remains from the Late Bronze Age have been found at the ancient tell, Dibon appears in a topographical list of pharaoh Tuthmosis III (1504–1450 b.c.e.).[45] From Dibon they proceeded to the rugged western escarpment of the Moabite plateau. On this plateau lies Mount Nebo, where Moses is supposed to have viewed the Holy Land before he died (in fact, he had died years earlier in the wilderness). Below the escarpment were the broad plains of the Dead Sea pull-apart basin into which the Jordan River flowed. Across the Jordan from the Israelites was Jericho, the gateway to central Canaan from the east.

Here the Israelites stopped, for an unknown amount of time, in what would become the tribal territory of Reuben. It seems they settled here with their flocks and probably set up a cult center at Shittim (see figures 9.1 and 9.2). But events occurring in the outside world would intrude upon them. Triggered by a series of natural events, profound political change followed by deadly plague would disrupt the Israelites' pastoral existence. Then, another, quite different, natural event would provide them with what to their eyes was a divine sign that they should cross the Jordan and settle once more in the land of Canaan.

CHAPTER SEVEN

Meanwhile, Back in Civilization

While the Israelites were spending their three to four generations wandering in the wilderness, other events were taking place that would eventually provide them with the opportunity to establish themselves once again in the land of Canaan. The trigger was a series of global and regional changes in the Earth's climate. These changes facilitated the emergence of a disease that in turn became linked to a series of events in Egypt. The consequences of these events proved far-reaching and eventually resulted in the end of the Middle Bronze Age in Canaan—and the return of the Israelites to their homeland.

REGIONAL AND GLOBAL CLIMATE CHANGES

Volcanic eruptions, especially large ones such as the Aniakchak and Minoan eruptions, will lower global temperatures for one to five years and, in the Middle East, result in cooler and wetter winters during those years.[1] In normally semiarid central and western Anatolia, this extra winter rainfall would have produced exceptionally bountiful harvests for several years, which in turn assisted the Hittites of central Anatolia in their expansion southward and eastward, and pushed other peoples, particularly the Hurrians, into Canaan (see chapter 9).[2]

Around the time of the Minoan eruption other longer-term and larger-scale climate changes were taking place. In Europe, the glaciers of the Swiss and Austrian Alps began to advance and the Alpine tree line started to retreat. This marked the beginning of a cold and wet phase that commenced about 3340 ± 100 radiocarbon years B.P. Other workers date the onset of this cold stage at 3440 ± 60 B.P. and its range at 3600–3200 radiocarbon years B.P.[3] This change to a colder and wetter climate is called the Löbben phase in the Swiss and Austrian Alps and the Pluvius phase in the French lake district. It surpassed the Little Ice Age (1300–1850 C.E.) in its intensity.[4] Tree-ring densities show an abrupt cooling starting in the later decades of the seventeenth century B.C.E., a recovery in the sixteenth and fifteenth centuries, and a renewed cold period from about 1350–1340 B.C.E. to about 1200 B.C.E.[5] This renewed cold spell coincided with a sunspot minimum that lasted from 1420 to 1260 B.C.E., much like the Spörer and Maunder sunspot minimums in the Little Ice Age.[6] Lake sediments and pollen cores from the Swiss and French Alpine lakes record much the same climatic situation, and they can also be dendrochronologically tied to calendar dates. Wood preserved in Swiss lake sediments shows that high water levels in the lakes began in the seventeenth century B.C.E., and that shortly before 1600 B.C.E. people stopped building lakeside dwellings because the water was too high. This phase ended 1100–1050 B.C.E., when lake levels dropped enough to allow lakeside building again.[7]

Beginning somewhere between 4,500 and 4,200 radiocarbon years B.P., central Africa became increasingly dryer and cooler. Across wide swathes of territory, tropical rainforests were replaced by seasonally dry forests and, especially in East Africa in the vicinity of Lake Victoria, by grasslands.[8] This change in climate was caused by a northward shift in the Intertropical Convergence Zone (ITCZ) which also shifted the monsoon rains northward, bringing increased moisture to areas across the southern Sahara and the northern parts of the Ethiopian plateau, beginning shortly before 3700 radiocarbon years B.P. and ending shortly before 3000 radiocarbon years B.P. (about 1250 B.C.E.).[9] This shift in rainfall brought increased moisture to the catchment area of the Blue Nile and its tributary the Atbara River, and was the cause of the very high Nile flood levels of Egypt's Middle Kingdom and the generally abundant Nile floods of the Eighteenth and early Nineteenth Dynasties.[10]

Microscopic pollen grains and diatoms from East African lake sediments record abrupt fluctuations between dryer and moister conditions starting about 7,200 radiocarbon years B.P. and lasting to about 2200 B.P. This was a time when climate became markedly more seasonal.[11] During this period in East Africa, there were frequent shifts between dry periods and times when rains brought dramatic increases in the plant cover. This sort of transition from dry to wet conditions has often been linked to outbreaks of disease, such as the deadly Ebola virus in central Africa, although the virus's animal vector remains unknown.[12]

The climate shifts in East Africa that brought increased moisture produced more vegetation and thus more food, allowing many rodent populations to explode. One of these, the multimammate mouse, is the principal host for fleas that carry the bubonic plague bacillus *Yersinia pestis*.[13] Cool (but not cold—fleas reproduce best between 18 and 27 degrees Centigrade) temperatures are also essential, because the plague bacillus will not block the foregut of an infected flea when the outside temperature is above 28 degrees Centigrade (about 80 degrees Fahrenheit). At higher temperatures, plague bacilli pass through the flea's digestive track. If it is cooler than 28 degrees Centigrade, an infected flea will be unable to ingest the blood it draws from the animal it bites. Instead, it regurgitates this blood, now mixed with some thousands of the plague bacilli, into the animal it is biting.[14] Most rodents (such as the multimammate mouse) that spend their lives harboring plague-carrying fleas are immune to the disease, but when their population expands rapidly, they are likely to come into contact with other rodents (such as rats) or predators that are not immune. If (or more likely when) these other animals acquire plague-carrying fleas from the first group, the new flea-carriers will fall ill of the disease and often die. Since a flea can survive unfed for one to three months, a plague-carrying flea will still be alive—and hungry—long after its new animal host has died. Given a chance, these fleas will abandon their dead host in search of a new source of blood, carrying the plague with them. In this way, plague-carrying fleas can reach human beings.[15]

When *Yersinia pestis* enters the human body from the bite of an infected flea, it multiplies in the body's lymph nodes. Two to six or eight days after the flea bite, the human victim gets a sudden fever, weakness, and headache. The lymph nodes swell, and one or more classic "buboes" form, usually in the groin or armpit nearest to the fleabite, and sometimes on the neck. These lumps are exquisitely painful and become filled with plague bacilli. In later stages the urine may turn red or "purple" with blood. Often there are skin rashes, as blood vessels rupture and the skin gets brownish or violet-colored, especially near the affected lymph nodes. These areas can ulcerate and result in gangrene. If the bacilli reach the lungs the plague becomes pneumonic; if it massively invades the blood it becomes septicemic. When untreated by modern antibiotics, 40%–60% of those affected by the bubonic form of the plague will die, while fatalities for the pneumonic and septicemic forms are virtually 100%. In the twentieth century, about 20%

of all plague cases were of the latter two forms; the percentage may have been higher in earlier centuries.[16]

The first widely recorded pandemic of plague occurred in the sixth century C.E. and is often referred to as the Justinian plague. It happened, not coincidentally, after a massive volcanic eruption in southeast Asia, possibly from the Krakatoa volcano, suddenly lowered world temperatures for several years.[17] In this case, the origin of the epidemic was the rodent reservoir in East Africa. The disease was spread by flea-carrying rats transported on ships from East African ports. These ships sailed from East Africa around the Horn of Africa, through the Gulf of Aden, up the Red Sea, and through a canal into the Mediterranean.[18] From there the infected rats moved to other ships and to cities throughout the Mediterranean world.

The spread of bubonic plague by shipborne rats and their fleas is not the only way the plague can travel. In the Middle Ages, the disease passed along the caravan routes of central Asia, and before that it was carried by Mongol horsemen from southern China and Burma to the central Asian steppes.[19] Many other species of flea besides the common rat flea *Xenopsylla cheopis* can carry the plague bacillus, and so can ticks and the human louse. Researchers (mostly French) who have studied the plague in the Middle East are convinced that, in past plagues in that area of the world, there has been a significant amount of human-to-human transmission when infected fleas or lice jumped from one person to another. Sometimes, even exposure to lice- or flea-infested cloth or clothing can spread the disease. Although a flea cannot become infected initially by a biting a person, it can get the infection from a sick mouse or rat, then jump from the rat to one human and to another, infecting each on its way. In premodern times, nearly all humans carried fleas and lice.[20]

Historical records in Europe, the Middle East, and along the southern margin of the Sahara (the Sahel) show that the bubonic plague most frequently occurred when the climate was cooler and moister than usual, including during the Little Ice Age.[21] The Löbben period was remarkably similar to the Little Ice Age, climatically. Plague would have been likely then. Moreover, plague outbreaks "usually follow military or commercial trade routes, and so may be either slow or fast depending on the prevailing political or social conditions."[22]

The ships that carried the plague bacillus from East Africa to the Mediterranean in the sixth century C.E. were principally carrying ivory.[23] Ivory has been one of the primary trade items from East Africa to the Mediterranean since pharaonic times. During Egypt's Second Intermediate Period, trade goods, including gold, ivory, ebony, and exotic animal skins, moved from sub-Saharan Africa north to Avaris and from there across the eastern Mediterranean. During the reigns of the later Hyksos rulers Khayam and Apophis (who apparently established a military presence along the entire length of the Egyptian Nile[24]), trade items from East Africa would have freely passed up the river from Nubia all the way to Avaris. It was at this time—during the reign of Seqenenre Tao of the Theban Seventeenth Dynasty—that there was said to have been plague in Thebes.[25] A later Egyptian historian, Hecataeus of Abdera (300 B.C.E.), wrote that the Egyptians interpreted this plague as the displeasure of their gods at alien rites and customs and so they expelled the Hyksos.[26]

THE END OF THE HYKSOS IN EGYPT AND THE SPREAD OF THE PLAGUE TO CANAAN

Seqenenre Tao was, nominally at least, a vassal of the Hyksos, and he seems to have died in battle with them—his mummified remains show the unmistakable marks of violent death by knives, clubs, and battleaxes, as well as a hasty embalming.[27] His successor, Khamose, initiated a military campaign against the Hyksos that reached the walls of Avaris itself, probably toward the end of Apophis' reign. Following his northern campaign, Khamose turned south to attack the Hyksos' allies, the Nubians of Kerma. Khamose closed the Nile to the Hyksos and forced them to communicate with their southern allies via the oases of the Western Desert.[28]

Khamose reigned for only a short time and was succeeded by his nephew Ahmose, Seqenenre's son (see table 7.1).[29] When Ahmose became old enough to lead an army, he too attacked Avaris. There is a much-debated text written in the eleventh year of an unnamed ruler that documents the movements of an Egyptian prince and his forces against Avaris. Donald Redford concludes—correctly, I think—that the papyrus is dated to the last Hyksos ruler, Khamudy.[30] It records the opening moves of Ahmose's campaign to defeat the Hyksos and drive them out of Egypt, which in the higher Egyptian chronology that the later seventeenth century B.C.E. Minoan eruption date requires, took place about 1550 B.C.E. (see table 7.1 for a possible reconstruction of the dating sequence).

Ahmose first took the fortress of Sile (Tell Hebua I—see figure 2.1) on the border between the Delta and the Sinai Peninsula, thus cutting Avaris off from land contact with the southern Canaanite cities and their food supplies. Then the Theban monarch cut the Hyksos capital off from the

TABLE 7.1.

Proposed Date Ranges for the Fifteenth, Seventeenth, and Early Eighteenth Dynasties

Hyksos XVth Dynasty		Egyptian XVIIth and XVIIIth Dynasties	
Reignal years		Reignal years	
B.C.E.	Ruler	B.C.E.	Ruler
1658–1639	Salitis/Sheshi		
1639–1602	second (Yʿakub-hr?), third (Khayan), and fourth (Ianassi?) rulers		
1602–1561	Apophis	–1574	Seqenenre Tao
		1574–1570	Kamose
1561–1550	Khamudy	1570–1546	Ahmose
		1546–1525	Amenophis I
		1525–1504	Tuthmosis I and II
		1504–1450[a]	Tuthmosis III
		1452–1426	Amenophis II

Notes: The dates for the Hyksos Dynasty were arrived at by using the 108 years given in the Turin papyrus for the XVth Dynasty and 1550 as the estimated date for its end. See Shaw, *Introduction: Chronologies and Cultural Change in Egypt*, 7. Salitis is given a reign of nineteen years following Manetho, Khamudy is given a reign of eleven years following the notation on the back of the Rhind mathematical papyrus, and Apophis' reign of nearly forty-one years is derived from the Turin papyrus (see text and accompanying references).

The dates for the XVIIIth Dynasty are based on a 1504 B.C.E. accession date for Tuthmosis III, which best fits the lunar calendar evidence: L. W. Casperson, "The Lunar Dates of Thutmose III," *JNES* 45 (1986): 139–50. The twenty-one years for Tuthmosis I and II were worked out by J. Darnell, cited in H. Goedicke, "The Chronology of the Thera/Santorin Explosion," *Ägypten und Levant* 3 (1992): 62: twenty-one years, one month, and eighteen days plus five epagomenal days. The reign lengths for Amenophis I and Ahmose are generally agreed upon as twenty-one and twenty-five years, respectively. Kamose is believed to have reigned only three or four years.

[a]Includes the reign of Hatshepsut.

sea by taking control of the feeder canal that brought the Pelusic branch of the Nile to the lake just north of Avaris, which served as the city's harbor. Finally, Ahmose attacked Avaris by land from the south.[31]

According to the third century B.C.E. Egyptian historian Manetho, the Egyptian pharaoh failed to take the Hyksos capitol by direct attack.[32] Instead he was forced to besiege it. Logistically speaking, sieges are among the most difficult of military operations, for they usually require massive amounts of food and other supplies to be transported to the besieging army (or navy). The Egyptians usually attempted to overcome this problem by investing a city just before its harvest, when stocks of grain in the town would be low, and their own armies would be able to live off the produce of the surrounding fields.[33] This was not the strategy followed at Avaris, however, because the Egyptian text states that the thrust against Sile occurred near the end of the first month of the inundation season, long after the end of the local harvest.[34]

Manetho relates how the Egyptian pharaoh (incorrectly called Thummosis) concluded an agreement with the Avarans that allowed them to leave overland, with their families and their possessions, across the Sinai Peninsula to Syria (Canaan). The one contemporary Egyptian account of the taking of Avaris speaks of looting it, but not of the forcible capture of the city. The archaeological evidence also suggests that there was no widespread destruction at the end of the Hyksos occupation.[35] Manetho and other late Egyptian historians equated the Avarans with the Jews, who in their own time occupied the land (Judea) the Avarans fled to (southern Canaan). According to another Egyptian historian, Apion, the Jews (that is, the Avarans) who left Avaris and crossed the Sinai all had buboes in their groins.[36] Buboes in the groin are, of course, a classic indicator of bubonic plague.

In the mid-sixteenth century B.C.E. Hearst Medical Papyrus and in the mid-fourteenth century B.C.E. London Medical Papyrus there are references to what is called the "Canaanite" or the "Asiatic" illness: "when the body is coal-black with charcoal (spots)," and "When the body is coal-black with charcoal (spots) in addition to the water (=urine) as red liquid (i.e., bloody). . . ." Hans Goedicke maintains that these are clear references to bubonic plague, and that, according to the Egyptians at least, the disease came from the Canaanites.[37] Goedicke also suggests that the incidence of plague in Canaan was the reason why Ahmose's son, Amenophis I, did not follow up his father's victory at the city of Sharuhen (the main Hyksos city in southern Canaan, taken by Ahmose a few years after Avaris fell—see figure 5.4) by invading the rest of Canaan. Amenophis I's successors, Tuthmosis I and Tuthmosis II, made only small forays into Canaan itself.[38]

When Ahmose besieged Avaris, he would have brought the Theban army's food up the Nile by boat from the south of Egypt and stored it where it was easily accessible for his troops. In this way rodents from the south, including some who may have come up from East Africa in trading ships carrying ivory, ebony, and gold, were transported on the pharaoh's grain ships north to the Delta. Once there, the rodents from the south would have mingled with the local Delta grass rat, whose native flea is probably *Xenopsylla cheopis*.[39] As the grain was consumed, the rodents would have searched for other food. Unlike the soldiers, the rodents would have had no trouble crossing the siege lines and walls to Avaris itself. There they would have come into direct contact with the besieged population. In Avaris, as in besieged cities throughout history, the rodents themselves may even have become food. In any case, the plague-carrying fleas would have spread effortlessly from rodent to human. An outbreak of deadly disease, a common consequence of military campaigns, and especially sieges, would have been a very good reason why the Egyptians allowed the Hyksos to leave Avaris unmolested.

Avarans fleeing across the northern Sinai to Sharuhen and other cities and towns of southern Canaan would have taken the plague with them. Plague-carrying fleas would have ridden on human bodies or in their clothes or on their animals, or on rodents hiding in any of the food brought along. The massive walls of the southern Canaanite cities and towns provided no protection against this microbial invasion. As people fled the contagion in one town by going to another, settlements throughout the area would have experienced outbreaks of this deadly disease.

Outbreaks of plague are usually episodic over an extended period of time, such as the repeated outbreaks in the Mediterranean world in the sixth and seventh centuries C.E.: "Sometimes it would spread to myriad towns and villages in a single year, while on other occasions it would bide its time, skulking in a few quiet or remote localities, only to burst forth from these nameless havens of death a few years later."[40] The sixteenth century B.C.E. plague likely followed a similar pattern.

A clear indication of a mid-sixteenth-century B.C.E. plague in southern Canaan is found in the tombs of Jericho. Here, in tombs used at the very end of the Middle Bronze Age just before the city was destroyed (Tomb Group V), there is evidence of an epidemic—approximately 53 bodies of all ages were buried at one time in six tombs and the tombs were never re-used, unlike other, earlier tombs.[41] Jericho was both connected to the coastal-to-inland trade network and a transfer point for overland caravans coming north from Arabia. At Jericho, trade items carried from the Mediterranean coast could be exchanged for aromatic spices, especially frankincense and myrrh, brought up from southern Arabia as caravans passed around the eastern flank of the Dead Sea and headed northward to Syria.[42] Thus it would have been a likely destination for travelers carrying the plague.

Indications of this plague also appear in several biblical passages. In Exodus 23:28–30 and Deuteronomy 7:20–22 there is the statement that the Lord will send hornets in front of the Israelites, among the peoples of Canaan, to drive them out little by little (in the course of a year). In written Hebrew the word for hornets closely resembles another word that was used for severe infectious diseases that afflict the skin. This similarity has caused commentators and scholars since the twelfth century to translate "hornets" in these passages as "pestilence."[43] Deuteronomy 7:15 says that the Israelites will not be affected by the dread diseases of Egypt they had previously experienced but that God will lay them upon those who hated the Israelites.

This reference to previous Egyptian diseases may harken back to the plague mentioned in Numbers 25:8–9, a plague that killed a good many Israelites (the text says 24,000, a typical exaggeration of a story passed orally). It is implied that the plague was caused by Israelite men having sexual relations with Moabite or Midianite women as part of the worship of the Canaanite god Baal of Peor,[44] when the Israelites were encamped just east of the Jordan River at Shittim, in what was to become the tribal territories of Reuben and Gad (see figures 9.1 and 9.2). In later times, Shittim was part of the Kingdom of Moab. In oral transmission the tellers of a story will change unfamiliar proper names to familiar ones. In this way the term "Midianite" would have been changed to "Moabite" because it made sense to Israelites in later times, just as the change from "Canaanite" to "Judean" made sense to Egyptians in later times in their stories about the defeat of the Hyksos.

Several scholars also argue for the antiquity of these text passages and the primacy of Midian in this story, including W. F. Albright, who suggested that the Midianites were donkey caravanners.[45] Midianite donkey caravanners could easily have acquired the plague at Jericho, brought there by fleeing "Egyptians"—that is, Hyksos; it could then have spread to the Israelites camped not far away. Afterward the Israelites avenged themselves on the Midianites. The ensuing slaughter of the Midianites described in Numbers 31 includes some rather remarkable precautions that the Israelites took after they had killed the Midianites and taken Midianite girls captive: "Camp outside the camp seven days; whoever of you has killed any person or touched a corpse, purify yourselves and your captives on the third and on the seventh day. You shall purify every garment, every article of skin, everything made of goats' hair and everything made of wood ... gold, silver, bronze, iron, tin, and lead—everything that can withstand fire, shall be passed through fire, and it shall be clean. Nevertheless it shall also be purified with the water for purification; and whatever cannot withstand fire shall be passed through the water. You must wash your clothes on the seventh day, and you shall be clean; afterward you may come into the camp" (Numbers 31:19–24). These strictures are, in fact, extreme simply for ritual purification, including as they do the washing or burning of personal possessions of the Midianites in addition to the remarkable (for that time) amount of cleansing of body and clothing. We should recall that the plague manifests itself in two to eight days and usually in two to six days ("wash on the third and the seventh days"), that both lice and fleas reside in one's clothing as well as on one's body, and that goats are carriers of the plague bacillus ("everything made of goats' hair").[46]

From roughly the same time period, the eighteenth century B.C.E. archives of Mari in northern Mesopotamia mention that King Zimri-Lim ordered the isolation (quarantine) of a woman who came down with skin lesions; even the personal possessions of the patient were to be avoided. Obviously, then, the ancients knew about the communicability of disease.[47] The steps taken in Mari are essentially the same, though far less extreme, as the purification rites of the Israelites after they had slaughtered the Midianites.

After the slaughter of the Midianites, the Israelites remained encamped on the east side of the Jordan River. At this time their leader was Joshua. It seems clear from the biblical passages that it was Joshua who urged the Israelites to cross the Jordan and encamp on the west side of the river. However, Joshua might not have been successful in getting his people to move but for another natural event that convinced the Israelites it was their god's wish that they should cross the Jordan River and conquer Jericho.

CHAPTER EIGHT

The Destruction of Jericho

THE ANCIENT TELL OF JERICHO

Since the nineteenth century the ancient tell of Jericho, now known as Tell es-Sultan, has been the focus of archaeological interest as excavators have attempted to find traces of the biblical account of the fall of Jericho at the site. The first of three archaeological excavations at Tell Jericho in the twentieth century was an Austro-German expedition led by biblical scholar Ernst Sellin and archaeologist Carl Watzinger in the years 1907–1909 and again in 1911.[1] They found what they believed were traces of the massive walls destroyed by the Israelites. Later, these walls proved to be from the end of the Middle Bronze Age, much too early, in the scholarly opinion of the time, to be linked to the biblical Israelites.[2] A second excavation team headed by British archeologist John Garstang dug at Jericho in the 1930s, and the third, and most famous, series of excavations was conducted by Kathleen Kenyon in the 1950s.[3] Kenyon traced human occupation at the site back more than 10,000 years, but she also confirmed that the walls found by Sellin and Watzinger were indeed from the end of the Middle Bronze Age.[4] Her work clearly showed that there had been no walls around Jericho in the Late Bronze Age, when the Israelite conquest was supposed to have occurred.

The various archaeological teams were able to trace the remains of both Early and Middle Bronze Age walls around the north, west, and south sides of the ancient tell (see figure 8.1). The east side had been cut through by a modern road, destroying whatever remained of the site's eastern walls. From the surviving remnants, it seems that both Early and Middle Bronze Age walls were built to include the town's permanent water source, the spring, within them.[5] In this way an attacking army could not cut the town off from its water source. More recent geological investigations in

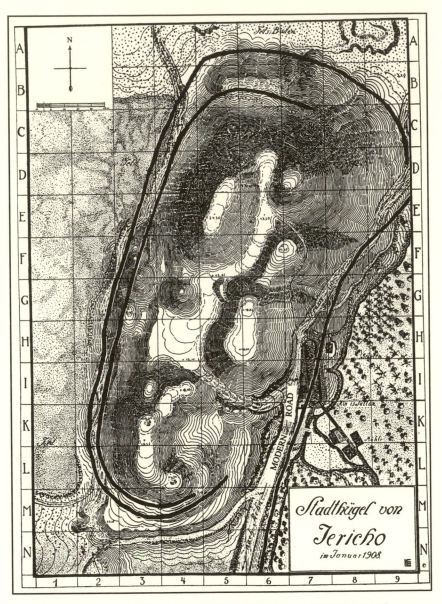

FIGURE 8.1. Plan of the tell of Jericho by Sellin and Watzinger in 1908 (Sellin and Watzinger, figure 1). Double dark lines encircling the south, west, and northern sides of the tell mark the locations of the Middle Bronze Age walls found by Kenyon (from Neev and Emery, figure 4.9). The dark line extending from E-9 south-southwest to N-6 is the approximate line of the fault. The lozenge-shaped feature in H-7 and I-7 is the spring. Most of the eastern edge of the tell was destroyed when a modern road was constructed in the nineteenth century.

the area, however, have shown that this apparent advantage also contained within it a fatal weakness.

A GEOLOGICALLY ACTIVE BASIN

Ancient Jericho is on the western edge of the Dead Sea pull-apart basin. The basin itself forms part of the boundary between the Sinai micro- or subplate and the Arabian tectonic plate.[6] The Dead Sea depression is, in fact, simply a continuation of the geologic rift that contains the Red Sea and the Gulf of Aqaba farther south. North of the Dead Sea depression, this rift stretches through the Jordan River Valley into Syria and northward to the boundary between the Arabian and the Eurasian plates in southern Turkey.

Because it is part of an active tectonic rift, the entire Dead Sea depression is filled with faults, both major and minor. The eastern edge of ancient Jericho is directly above one active north-south normal fault. Groundwater from deep in the rocks of the adjacent escarpment seeps up through the fault and forms a perpetual spring, the Spring of Elisha (see figure 8.1).[7] This permanent source of fresh water is the reason that Jericho has been occupied for thousands of years, since the end of the last Ice Age.

The disadvantage to extending the eastern town wall around the spring was that it would cross the fault in several places. As a consequence, the eastern wall was repeatedly damaged or destroyed as the ground moved along the fault line. Substantial earthquakes would have caused even greater damage to both the town and the walls.

THE BRONZE AGE WALLS OF JERICHO

The Early Bronze Age wall, on the summit of the tell, was built of free-standing sun-dried bricks. With nothing to support it, the wall collapsed a number of times from earthquakes; in the thousand years between 3100 and 2100 B.C.E., the Jericho walls were destroyed about seventeen times.[8] Toward the end of the Early Bronze Age the town was destroyed, burned, and abandoned for a time. Later, more—and perhaps different—people came and built houses of green mud bricks on top of the tell, but again there was an earthquake, fire, and abandonment. This was followed by a period of major erosion in the first part of the Middle Bronze Age. Earth, brick fragments, and other occupation debris from the top of the tell eroded down the slopes.[9]

FIGURE 8.2. Photo of the remains of the Middle Bronze Age walls of Jericho excavated by Sellin and Watzinger (p. 48, figure 26). Note the buildings erected on the sloping area between the outer wall (man at the top of the wall provides scale) and the remnants of the inner wall on the top left. These dwellings fit the description of Rahab's house built into the city (that is, the outer) wall with windows in the outer wall. Window embrasures in the outer wall, as mentioned in Joshua 2:15, can also be seen.

Eventually, in the later Middle Bronze Age (MB IIB), these eroded sediments were used to form the rampart for another, far more impressive wall. This wall, built of stones, started at the base of the tell and rose some 4.5 to 5.4 meters (over fifteen feet). It is called a revetment wall and is similar to many others that surrounded Middle Bronze Age cities in Canaan. In front of the revetment was a sloping ramp made of crushed stones. Behind the revetment wall the massive earthen rampart, more than twenty meters wide, sloped up to the upper part of the tell. Buildings were built upon the rampart, sloping up to the upper regions of the tell (see figure 8.2). There was also a second wall crowning the plastered slope of the earthen rampart, similar to other Middle Bronze Age city walls.[10]

On top of the stone revetment wall was another wall, a parapet wall, of sun-dried mud brick. It was at least eight feet (about 2.4 meters) high and had windows along its outer surface. These windows, as the pictures of Sellin and Watzinger clearly show, belonged to buildings built out of the mud-brick wall that stood on top of the revetment wall. Unlike the revetment wall, this mud brick wall was, like the Early Bronze Age mud brick walls, unsupported and susceptible to collapse in the event of a major earthquake. In fact, mud brick is one of the most susceptible of building materials to earthquake damage.[11]

MIDDLE BRONZE AGE DESTRUCTION

The Middle Bronze Age city of Jericho came to a violent and fiery end in the mid-sixteenth century B.C.E. The archaeological estimate for this destruction is about 1550 B.C.E.; one recent radiocarbon estimate is 1571–1529 B.C.E., a remarkable agreement.[12] The city had been thoroughly burned. Kathleen Kenyon found evidence of fire covering the whole area of her excavations, about 52 by 22 meters. The tops of the wall stumps were covered by a burned debris layer about a meter thick. Upper stories of buildings had collapsed into lower floors, and walls and floors were hardened and blackened.[13]

One wonders how the fire happened. One unusual feature of the MB destruction layer of Jericho was the presence of large storage jars filled with grain. Obviously the city had not been under siege for any length of time, since so much food was found in the city.[14] Nonetheless, the stores of grain would have provided ideal tinder.

Could an earthquake have caused the fire? In modern times fires usually follow earthquakes, as gas mains rupture and electrical cables break. But even before gas and electricity, earthquakes caused fires. Immediately following the massive 1755 earthquake in Lisbon, Portugal, fires from upset cooking fires and fallen oil lamps that ignited stores of wood and thatch did even more destruction than the quake itself.[15] Kathleen Kenyon noted that the eastern walls of certain rooms at Jericho had collapsed before they were affected by the fire.[16] This suggests an earthquake, perhaps one involving the fault on the east side of the city.

THE BIBLICAL ACCOUNT IN JOSHUA

According to the biblical account in the book of Joshua, Joshua sent two spies to Jericho while the Israelites were encamped on the west side of the Jordan River. The spies spent the night with Rahab the prostitute, who hid them "with the stalks of flax that she had laid out on the roof" (Joshua 2:6).

Just after it is harvested, flax needs to be laid out and retted before it is processed into linen. Joshua 3:15 also puts the time of year at the harvest: "now the Jordan overflows all its banks throughout the time of harvest." Both these statements are in accord with the large amount of grain found in the town by the archaeologists. In return for hiding the spies, Rahab was promised safety for herself and all of her family, as long as she put a crimson cord in the window through which she let the spies down. "Then she let them down by a rope through the window, for her house was on the outer side of the city wall and she resided within the wall itself" (Joshua 2:15). A look at the excavations of Sellin and Watzinger (figure 8.2) shows how closely the Middle Bronze Age wall fits the description in Joshua 2:15. What Kenyon called the upper wall on the crest of the rampart is also an inner wall, and the parapet wall atop the stone revetment is the outer wall.

The description of the crimson cord hanging from the window on the town wall points not simply to a promise of safety, but to a stratagem for gaining access to the town. Rahab's house was probably close to the town gate on the east wall, to better welcome her potential clients. A picked force of Israelites, entering at night through her window (marked by the cord) up the rope that the spies used, could have rushed to the gatehouse, overcome the guards, opened the gate, let in the rest of the Israelite fighters, and taken the town.

After hiding for three days the spies crossed the Jordan River and reported back to Joshua at Shittim. This is the same Shittim mentioned in Numbers 25:1 where plague devastated the Israelites (see chapters 6 and 7). Numbers 14:37 reports that all the spies save Caleb and Joshua who had been sent out in the earlier reconnaissance of Canaan (just after the Exodus) died of plague. It would certainly have made sense to later Israelite storytellers for the unfaithful spies in Numbers 14:37 to have died of the plague, but it is more likely that it was these later spies, from Rahab's time, who succumbed to the disease.

Acting on the spies' report, Joshua and the rest of the Israelites moved to the edge of the overflowing Jordan River and prepared to cross. Joshua told the people: "the waters of the Jordan flowing from above shall be cut off; they shall stand in a single heap . . . [and] the waters flowing from above stood still, rising up in a single heap far off at Adam, the city that is beside Zarethan, while those flowing toward the sea of the Arabah, the Dead Sea, were wholly cut off. Then the people crossed over opposite Jericho" (Joshua 3:13, 16).

Earthquakes of varying magnitudes are common along the faults of the Dead Sea depression. Earthquakes have long been measured by the Richter "local magnitude" scale (M_L), but seismologists also use another type of scale, called the Modified Mercalli Intensity Scale. This measures how intense the shaking is at any given spot.[17] It is particularly useful because it describes the sorts of damage different intensities will produce. For example, with an earthquake of Mercalli Intensity VII (estimated M_L 5.5–6.1), poorly built structures break, brick chimneys break at the base, sand and gravel banks cave in. Walls are damaged, but they probably do not collapse.

Studies by Israeli geologists have shown that earthquakes with a local magnitude (M_L) greater than 5.5 on the Richter scale occur about once every six hundred years near Jericho.[18] But they don't seem to occur at evenly divided intervals; in fact, like earthquakes in many other parts of the world, the earthquakes along the Dead Sea Fault tend to cluster. Three earthquakes with an estimated magnitude of 6.2 or more occurred within a hundred years in the Dead Sea–Jordan River fault zone in the nineteenth and twentieth centuries, in 1834, 1837, and 1927. The most recent of these, on July 11, 1927, was recorded on modern seismographs.[19]

Northward of the town of Adam (which is about twenty-eight kilometers north of Jericho), the Jordan River flows for about twenty kilometers between high cliffs composed of soft marls of the Lisan Formation.[20] Earthquake-caused landslides from these cliffs have dammed the flow of the Jordan River repeatedly in the past. After the 1927 earthquake, the river was cut off for twenty-two hours. In 1546, landslides from a large earthquake caused an identical stoppage for two days. In 1267 another earthquake caused stoppage from midnight until 10:00 the following morning, after which the bridge of Damieh had to be repaired.[21] These events are identical to the scene described in Joshua 3. The stoppage of the river in Joshua's time seems to have lasted for less than a day, similar to the duration of the 1267 earthquake.

DESTRUCTION OF JERICHO'S WALLS

After crossing the Jordan riverbed the Israelites set up camp at Gilgal on the west side of the Jordan. There they kept their covenant renewal feast (called the Passover in the biblical text) and "ate the produce of the land" (Joshua 5:11)—that is, they harvested what crops had not yet been brought in by the people of Jericho. The biblical text implies that the Israelite appropriation of Jericho's harvest and the siege of the town directly followed the Passover celebration. However, both barley and wheat were found in the storage jars that archaeologists recovered from the destruction of the town.[22] Apparently the Israelites interrupted the Jericho wheat harvest, which took place in May, after the barley had been harvested.[23]

The people of Jericho were now besieged. During the next few days, according to the book of Joshua, the Israelites marched around the city with warriors, priests, ark, and ram's horns. In Joshua 6 there is a good deal of confusion about the trumpets, that is, the ram's horns. The order given in verse 10 to keep silent until a war cry is raised does not agree with the many occasions when the trumpets are blown, sometimes by the soldiers and sometimes by the priests. In verses 14 and 15 one circuit of the town was made on the first six days and seven on the seventh day. The priests blow the ram's horns on each circuit on the seventh day (verse 4) or only on the seventh circuit (verses 5 and 16) or during all the circuits during all seven days (verses 8, 9, and 13). On the seventh day (seven is another ritual or perfect number), the Israelites blew their ram's horns, gave a great shout, and the wall fell down flat.

Trumpets and shouting do not bring down walls, but earthquake aftershocks can, especially since the mud-brick walls on both the parapet and the upper walls of the town would have sustained structural damage from the first quake. Almost all large earthquakes have aftershocks, the largest and most substantial ones usually in the month following the main quake.

EARTHQUAKE WAVES AND BUILDING COLLAPSE

Earthquakes along faults produce two types of seismic waves, (1) body waves (primary [P] and secondary [S]) from deep in the earth, and (2) surface waves. Body waves travel from the earthquake's epicenter at higher frequencies and speeds than surface waves; primary (P) body waves at higher frequencies can be heard by the human ear, often allowing people to hear an earthquake before they feel shaking from the surface waves.[24] There is thus a good possibility that it wasn't ram's horns that everybody heard, just before the walls of Jericho started to fall.

Buildings and walls have natural frequencies. Low buildings have higher natural frequencies than tall buildings. Sometimes the frequencies of earthquake waves match the natural frequencies of the buildings and walls they pass under. When this happens, a great deal of damage is done to the buildings and walls. A higher-frequency seismic body wave, one that can be heard, is more likely to match the natural frequency of the low buildings and structures such as those at Jericho.[25]

In front of the Jericho revetment wall (and on top of the crushed stone piled deliberately against it) Kathleen Kenyon found piles of red mud bricks "piling nearly to the top of the revetment."[26] These bricks apparently came from the parapet wall or from the upper wall of the MB city. Given Jericho's location, its eastern wall built on top of an active fault and surrounded by many other active faults, an earthquake or aftershock is the most likely cause of the collapse of the red mud-brick walls.

ISRAELITE CONQUEST OF JERICHO

Joshua 6:20 relates what happened next: "So the people charged straight ahead into the city and captured it." It has been suggested that they clambered up the piles of downed brick to enter the city.[27] Although the inhabitants were slaughtered, Rahab and her family were brought out, items of precious metals were taken, and the rest of the city, according to Joshua 6:24, was burned. Fires would have already started from the earthquake's aftershock; unchecked, they completed the destruction of the town.

Supposedly Jericho's oxen, sheep, and donkeys were devoted to destruction (Joshua 6:17–18): "The city and all that is in it shall be devoted to the Lord for destruction. . . . keep away from the things devoted to destruction so as not to covet and take any of the devoted things and make the camp of Israel an object for destruction, bringing trouble upon it." Could the recent plague in the town have anything to do with making most things in the city objects of destruction? During the Israelite attack on Jericho, Rahab and her family were brought from the city and set outside the Israelite camp—in effect, quarantined there, just as King Zimri-Lin of Mari required the quarantine of the woman affected with skin lesions.[28] Rahab and her family were said to have lived among the Israelites "until the present day" (Joshua 6:25). In fact, there is no evidence for anyone living at Jericho between the MB destruction and the middle of the Late Bronze Age. In the late fourteenth century B.C.E. a single dwelling structure and its outbuildings occupied one edge of the tell.[29] Although it is possible that the residents of this structure were or claimed to be descendants of Rahab, the story of her survival may simply have developed in the Late Bronze Age to explain this single residence on an otherwise deserted mound.

CHAPTER NINE

The Conquest and Settlement of Canaan

After the destruction of Jericho, Joshua returned to his base at Gilgal not far to the north of the ruined town and pondered what to do next. The plains and broad alluvial valleys of Canaan were heavily populated and still contained the fortified Middle Bronze Age cities. Because of this the hill country was to be preferred; the Canaanites' chariots could not be used there, and the steep terrain precluded deployment of heavily armed troops. A gifted military commander, Joshua realized that the terrain of the hill country worked to the advantage of his lightly armed and more mobile Israelite warriors.[1] His options, and those of his people, were limited by the nature of the land itself.

THE LAND OF CANAAN

Canaan, from Dan in the north to Beer-sheba in the south, stretches for about 220 kilometers (about 140 miles) (figure 9.1); south of Beer-sheba the Negev desert extends for another 190 kilometers to the Gulf of Aqaba (see figure 5.4). The distance from the Mediterranean to the Jordan river is only about eighty kilometers (fifty miles), from the flat coastal plain up through the foothills (the Shephelah—see figure 9.2) to the rugged highlands and down to the desert just east of the Jordan Valley and Dead Sea rifts. Rainfall is greater in the highlands, but the highland soils—the red terra rossas and the brown forest soils and rendzinas—are often shallow and rocky, better for pasturage in many areas than for arable farming.[2]

Farthest north in the highlands or hill country is Upper Galilee, a land of steep relief with high peaks, including the highest mountain in the area, Har Maron (1,208 meters). Southward, in Lower Galilee, the mountain

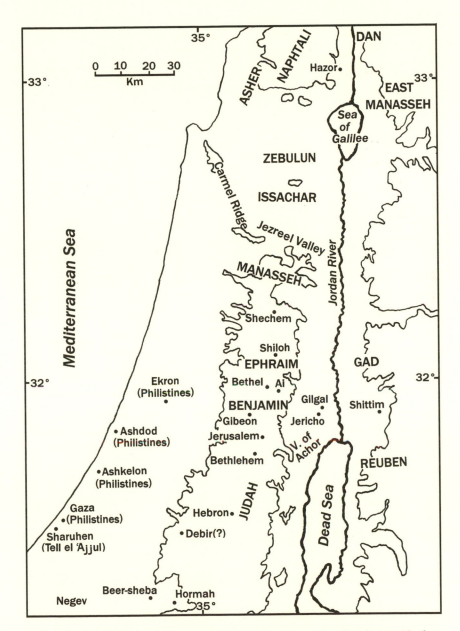

FIGURE 9.1. Late Bronze Age Canaan with sites and towns mentioned in the text. The four hundred meter contour line is shown. Names of the Israelite tribes are in capital letters.

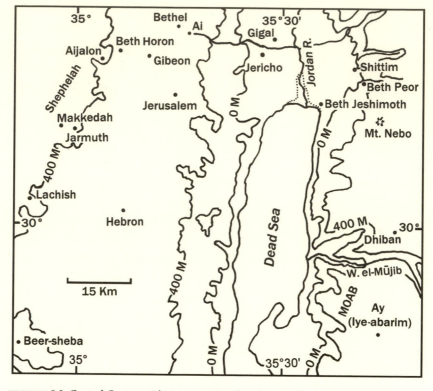

FIGURE 9.2. Central Canaan with sites mentioned in the Israelites' sixteenth century invasion. Dotted lines at the north end of the Dead Sea indicate the approximate extent of the northern bay that would have existed with a Dead Sea level of 375 meters below sea level (mbsl) or higher (see Frumkin and Elitzur, figure 5 and also text in chapter 9).

ridges are lower and separated by several east-west-trending valleys. Lower Galilee is divided from the central highland hills by the Jezreel Valley, which contains the most important east-west route through Canaan. South of the Jezreel are the low chalky hills and intersecting valleys of Manasseh, bordered on the south by Shechem.

South of Shechem is a rugged highland area, formerly heavily forested, with few inner valleys—the hill country of Ephraim and Benjamin. Ephraim has rugged limestone bedrock and lacks the broad plateaus of Benjamin and Judah. It is the most inaccessible part of the hill country. To the south, the saddle of Jerusalem separates Ephraim and Benjamin from the Hebron hills. Another important east-west route leads from the coastal plain through the Judean hills to Jerusalem, then down into the desert to Jericho and across the Jordan to the Transjordan plateau.

According to Egyptian sources, Jerusalem was the most important town in the southern hill region at that time.[3] Rather than take his chances against this powerful city or its territory, Joshua sent spies to reconnoiter a more northerly route up the wadis of the Judean watershed to the fortified town of Bethel, which commanded the ascent to the watershed (figure 9.2). Joshua's spies discovered that Ai, not Bethel itself, guarded the ascent at the edge of the watershed,[4] and that only a small force would be needed to attack it. The original Hebrew meaning of Ai is not "the ruin" as was once thought, but "the extreme limit"[5]—a good description of Ai's strategic location at the edge or limit of the watershed and the cultivable land.

Ai has long presented a serious problem in biblical archaeology. Despite the fact that the narrative in Joshua 7–8 contains graphic—and accurate—descriptions of the local terrain as well as a highly reasonable account of the military action, the archaeological picture reveals no occupation at the proposed site of Ai from the end of the Early Bronze Age (EB III: 2550–2350 B.C.E.) to the beginning of the Iron Age (Iron I: 1200 or 1220–1050 B.C.E.). Furthermore, the site was abandoned not long after 1050 B.C.E.[6] One attempt to solve this problem, by relocating Bethel and Ai, has met with little acceptance and has degenerated into a squabble over the location and significance of certain Roman milestones.[7]

The archaeological excavations at the generally-agreed-upon site of Ai (et-Tell—see figures 9.1 and 9.2) show that the Iron Age village was built within the upper reaches of the earlier, well-fortified Early Bronze Age town. The Iron Age occupants of the site took advantage of the still extant EB fortifications located to the northwest, west, and southwest, and of the EB walls on the north and south, and on the east where there had been a major gate. Even in the twentieth century, part of the EB walls were standing to a height of seven meters (well over twenty feet) and the remains of a city gate were still apparent. At the western edge of Ai a narrow saddle of land connects it to a series of rises between the town and Bethel itself. The high ground of these rises visually screens Ai from Bethel.[8]

Ziony Zevit has suggested that the two battles for Ai described in Joshua 7 and 8 were invented sometime after the Iron Age occupation of the town, possibly to "explain" the ruins. In the first battle the small Israelite force is defeated by the men of Ai who kill 36 of the attackers (Joshua 7:5a). Zevit translates the Hebrew word "sebarim" in the next sentence (Joshua 7:5b) as a ruin: "They [the men of Ai] pursued them [the Israelites] in front of the gate to the ruined walls"—that is, the remaining EB walls of the town—and down the slope.[9]

After this defeat, Joshua 7:7, 9 describes the Israelites' acute vulnerability among the far more numerous peoples surrounding them, another piece of evidence that they were not a "great host." Instead, they fear they will be surrounded and destroyed. To preclude this dire threat, Joshua decides to launch another, and better-planned, attack on Ai. First, he sends a small force (an ambush party of "thirty thousand"—probably one hundred or fewer men) by night to hide in the hilly rises that border the narrow saddle of land to the west of Ai. Joshua and the main force (probably several hundred men) then come up under the cover of darkness and camp across a ravine just north of Ai. He also sends a small screening force to block the approaches from Bethel, or to at least to give warning should reinforcements be sent from Bethel to Ai.

Early the next morning the king of Ai and "all the inhabitants of the city" went out and met the Israelites who then pretended to flee. Thus encouraged, the people of Ai pursued them. The story relates next that Joshua went up to the top of the slope and pointed, waved, or flashed his sword (or dagger) to signal the ambush party hiding to the west of Ai.[10] At this prearranged signal the ambush party rushed into the city, setting fires, their signal to Joshua that they had entered Ai. At this Joshua and "all Israel" turned back from their pretended flight and "struck down the men of Ai" (Joshua 8:21). After the inhabitants were slaughtered the town was looted and burned, and the king of Ai was hung upon a tree, his body thrown down at the city gate.

The degree of realism in this story is so marked that efforts to suggest it was "invented" in the Iron Age to explain a set of ruins are more than a little lame. Two Israeli military experts, Chaim Herzog and Mordechai Gichon, had another suggestion. If so many of the Early Bronze Age fortifications were still standing in the Iron Age, even more would have existed several hundred years earlier, at the end of the Middle Bronze Age. Expressing their appreciation of "the great strength of ruins prepared as defensive positions," Herzog and Gichon suggested that a contingent of people from Bethel, in order to forestall an attack on their town, occupied Ai as a fortified outpost once they heard of the Israelite conquest and destruction of Jericho.[11] Such a short-term occupation would probably leave no archaeological trace, especially since nearly all the fighting took place outside the remains of the town's walls. Even the intentional setting fire to the city (Joshua 8:19) might really have been only the lighting of a signal fire or two to let Joshua and the other Israelites know that the detached force had entered the ruins of the town proper. Upon seeing the rising smoke, Joshua and his force at once turned around and reversed their retreat while the picked men came out of Ai to surround the enemy and kill them. In typical oral exaggeration, the Israelites promoted the executed enemy leader to "king of Ai."

None of the biblical descriptions of the encounters at Ai—the initial confidence of the spies reconnoitering the area, the men and people of Ai running out after the Israelites on two separate occasions, the Israelite picked force entering the town and passing rapidly and easily through it to catch the people of Ai from their rear—suggests a walled and securely gated town with a settled population and intact buildings that would have prevented such easy movement. According to Joshua 8:25, only men and women composed the people of Ai—no children. Although women would likely have been at the outpost to cook and tend to various domestic needs, they would have left their children within the safer confines of Bethel. If there is an invented part of the story, it is the statement in Joshua 8:28 that Joshua burned Ai and made it a heap of ruins. It already was a ruin.

THE ISRAELITE-GIBEONITE ALLIANCE AND BATTLES AGAINST THE CANAANITES

After the battles at Ai the Israelites return to their base camp at Gilgal where they make a peace treaty with the four cities of the Gibeonites: Gibeon, Chephirah, Beeroth, and Kiriath-jearim. This alliance with the Gibeonites (for it was probably *was* an alliance) gave the Israelites control of the roads leading up from the foothills of the Shephelah to the highlands, especially the Beth-horon road.[12]

According to Joshua 10, the king of Jerusalem, upon hearing of the treaty, joined with the kings of Hebron, Jarmuth, Lachish, and Eglon and besieged the Gibeonites. Jerusalem was the dominant town in the southern hill country and had a substantial interest in the east-west route that passed through it down to Jericho, a route the Israelites now blocked. It would make a good deal of sense for Jerusalem's ruler to muster military contingents from some of his allies (or subordinate rulers), to move against an Israelite-Gibeonite alliance. The Gibeonites appealed for help to *their* allies, the Israelites. Joshua and his men launched a surprise attack at dawn and defeated the Canaanite forces, chasing them down the defiles of the Beth-horon pass and into the Aijalon Valley (see figure 9.2). Herzog and

Gichon suggest that early morning fog, common in the Aijalon Valley, played an important role in the battle, allowing the more lightly armed Israelites to defeat their enemy with help from the local Gibeonites, who rolled stones down on the fleeing Canaanites.[13]

This military action brought the Israelites out of the hill country and into the foothills or the Shephelah. The hills of the Shephelah are made of Eocene limestone, separated from the Judean highlands by a narrow "trough" of exposed Senonian chalk. The Israelites (probably with some of their Gibeonite allies) followed up their pursuit to prevent the enemy from reaching friendly fortified towns: "pursue your enemies, and attack them from the rear. Do not let them enter their towns" (Joshua 10:19). Only the remnants of the Canaanite force reached the town of Azekah by way of the Valley of Elah.[14]

The next part of the account is somewhat confused, but the Canaanite forces seemed to have fled south down the "trough" to the town of Makkedah. The Israelites, probably with their Gibeonite allies, followed and, supposedly, trapped the five kings in a cave. More likely the Israelites left a force to keep the Canaanites holed up in Makkedah while another group went on to wipe out whatever survivors they could find (Joshua 10:20). After this, "all the people returned safe to Joshua in the camp at Makkedah; no one dared to speak against any of the Israelites" (Joshua 10:21). This last phrase again suggests that the Israelites were only one part of the attacking alliance or coalition.

Joshua 10 goes on to relate how Joshua killed the five kings of Jerusalem, Hebron, Lachish, Jarmuth, and Eglon and then took Makkedah (see figure 9.2). Following this, he and the Israelites captured the cities of Libnah, Lachish, Hebron, Eglon, and Debir. Joshua 11 relates how, after returning to Gilgal, the Israelites turned north to defeat a consortium of northern kings at the battle of the waters of Merom and then burned Hazor. Joshua 12 follows with a list of the kings defeated by the Israelites under Joshua. In contrast, other parts of Joshua and Judges give different and sometimes contradictory stories of these conquests. Most interesting is a list in of the towns that the various Israelite tribes did *not* drive out or conquer.[15] Some of these towns are on the conquered kings list in Joshua 12.

Could the relatively small number of Israelite warriors, even with their coalition allies, have successfully assaulted the highly fortified Middle Bronze Age Canaanite cities? Would the ruler of Hazor, so far north of the Israelite base camp at Gilgal, even bother with the incursions of such a small group, let alone form a coalition against them? The king of Hazor in this story has the same name as the king of Hazor in a later story,[16] in which the army of Jabin of Canaan (or Hazor) is destroyed by the forces of Deborah and Barak. A King Jabon of Qishon is mentioned in an Egyptian thirteenth century B.C.E. topographical list, and Qishon is also mentioned in Judges 5, while archaeological evidence shows that Hazor was destroyed in the thirteenth century B.C.E.[17] The battle and subsequent destruction of Hazor in Joshua 11:1–13 is most likely a thirteenth century B.C.E. story that got included with the Conquest, as did other stories of later conquests by the Israelites. These are more examples of the "lightning rod effect" in oral traditions, when significant foundational events and leaders (such as the Conquest and Joshua) attract unrelated events from other time periods.

Another story that made its way into the Conquest traditions is from an earlier time period. In Judges 1:4–10, Judah and Simeon defeat Adoni-bezek at Bezek and go on to take Jerusalem and Hebron. Bezek is in the territory of Manasseh north of Shechem, and Adoni-bezek means "lord of Bezek," but the defeated king was more likely the ruler of Shechem.[18] If so, this story hearkens back to the story of the killing of Hamor and his son Shechem by Simeon and Levi in Genesis 34. In fact, both stories are probably two highly altered oral traditions of a single battle fought at Bezek between Jacob and his familial band on the one hand and the Shechemites on the other. This battle probably relates back to Genesis 48:22, which suggests (contrary to the versions in Genesis 33:18–19 and Joshua 24:32) that Jacob conquered Shechem. These are all indications that the story in Judges 1:6 was originally an earlier story from the time of Jacob, not a story from the Conquest period. The taking of Jerusalem and its king Adoni-zadek in Joshua 10:1, 3, 23–26 is not historical and results from the similarity of the two kings' names: Adoni-bezek and Adoni-zadek.[19]

THE DESTRUCTION OF CANAAN AT THE END OF THE MIDDLE BRONZE AGE

The archaeological record presents an even greater puzzle than the biblical text, for not only were all the identifiable towns in Joshua 10–11 destroyed or abandoned at the end of the Middle Bronze Age, but so were a great many more. In fact, there was wholesale destruction of nearly all the towns and villages in Canaan starting in the mid-sixteenth century B.C.E. Of the 249 recorded Middle Bronze Age sites in the central hill country, only twenty-seven survived into Late Bronze II in the hills of Manasseh, only five in Ephraim, and only one (Jerusalem) in the hill country of Benjamin. Only three of eight settlements survived in the hill country of Judah. In the coastal areas and northern regions the destruction or abandonment was less severe, but it still approached 60–65%.[20] At the height of the Middle Bronze Age, the estimated population of Canaan was 140,000; in the succeeding Late Bronze Age it was less than half as large, about 60,000–70,000 people.[21] What then happened to all these MB towns, villages, and cities and their inhabitants?

The long-favored explanation was that the Egyptians under the Pharaoh Ahmose and his son Amenophis I conducted campaigns of destruction throughout the country after destroying the main Hyksos center of Sharuhen.[22] It has even been suggested that many of the people of the central hill country were killed or sent to Egypt in mass deportations.[23] The problem with this explanation is that there is virtually no Egyptian textual evidence to support military activity in Canaan on such a massive scale during this time, and later Egyptian pharaohs did not usually destroy sites; they preferred to extract tribute from cities, not demolish them. In fact, the destructions and abandonments in Canaan seem to have taken place over the course of the whole sixteenth century B.C.E., with the inland regions being ravaged first and the coastal settlements and major valleys slightly later, the exact opposite of what the Egyptians would have done.[24]

A second explanation is that the Hyksos, escaping from Avaris and Sharuhen and fleeing to other cities and towns, intensified population pressure and intercity warfare throughout Canaan.[25] The conquest of Avaris and Sharuhen would have ended the trade network that supported most of the people in the hill country, leaving this densely populated area with no market for their grain, olive oil, or wine. But many of the smaller sites in the hill country had already been abandoned in the final part of the Middle Bronze Age during the Hyksos dynasty; their populations seem to have been drawn to the large cities of the southern coastal plain or strongholds in the highlands.[26]

Archaeologist Israel Finkelstein has proposed that widespread social breakdown caused much of the population of the central hill country to became nomadic pastoralists, and thus archaeologically invisible.[27] There is evidence for a sizable pastoralist population in the hill country in the Late Bronze Age (see below), but were these pastoralists descendants of the Middle Bronze Age populations of the area—or newcomers, such as the Israelites?

Nadav Na'aman of Tel Aviv University believes that the Hurrians, northern peoples from Anatolia, migrated into Canaan in large numbers during the late seventeenth and sixteenth centuries and started the chain of events that ended with so many large and small sites destroyed and abandoned. Egyptian textual evidence shows a marked increase in Hurrian names in Canaan from the seventeenth to the fifteenth centuries B.C.E.; unfortunately, there is no textual evidence from the sixteenth century B.C.E. However, these textual sources seem to indicate that the Hurrians were partial to cities, not the hinterland of the central hill country.[28]

There are also natural phenomena explanations: plague, earthquake, and fire. The most ambitious of these is the hypothesis of Neev, Bakler, and Emery, who maintain that the Mediterranean coast of the Sinai and Israel has been subject to fault activity and earthquakes for many thousands of years. According to them, the coast has risen and fallen three or four times in the last four thousand years; one tectonic oscillation occurred at the break between the Middle and the Late Bronze Ages, and another between the Late Bronze and the Iron Ages. They maintain that the sixteenth century B.C.E. transition from the Middle to the Late Bronze Age was marked by "catastrophic tectonism" and a change to a more humid climate, with more swamps and sand dunes near the coast.[29] These ideas are controversial, however,[30] and such *coastal* oscillations do not explain the destruction of the highland towns and villages.

And, finally, there is the "no real cause" explanation: Middle Bronze Age urban culture in Canaan was a self-organizing open complex system in which a period of stability was followed by short interval of strong fluctuation or chaos, characterized by nomadization and migrations, from which a new level of stability emerged—the Late Bronze Age urban culture. In this systemic approach, a small random, minor cause may have triggered the collapse of the entire network of Middle Bronze Age settlement.[31]

All of these possibilities do suggest a multifaceted picture of collapse at the end of the Middle Bronze Age. The fall of Roman Britain in the fifth and the first part of the sixth centuries of the Common Era serves as a useful comparison. In 410 C.E. the Britons were cut off from their political and military capital, Rome, due to a barbarian invasion on the continent. At the end of the Middle Bronze Age, the cities of Canaan were cut off from their trade centers of Avaris and Sharuhen, both conquered by the Egyptians. In Britain, barbarian invaders soon came from the north (the Picts), the west (the Irish to Wales), and across the North Sea (the Angles, Saxons, and Jutes). In Canaan, Egyptians penetrated as far as Sharuhen and Gaza in the west, Hurrians came from the north, Israelites from the east, and Kenizzites and others from the south. Native British efforts to repel their invaders were partially successful up to the middle of the sixth century, but then bubonic plague cut a swathe through British communities, leaving the Germanic invaders relatively unscathed.[32] In Canaan at the end of the Middle Bronze Age, although the effects and extent of the plague are unknown outside of Jericho, the disease must have decreased urban populations and left them vulnerable.

Na'aman, writing of the Hurrians, says, "the newcomers gradually sacked and ruined towns and villages in the inner parts of Canaan, blocked the roads and disrupted trade, despoiled the crops in the fields and finally conquered and destroyed major Canaanite centres."[33] Undoubtedly the invading Israelites also did some of these things as they penetrated and settled the central hill country north of Jerusalem, an area that had probably already lost a good deal of its population to the cities. It is not coincidental that, of all the areas of Canaan, the hill country of Benjamin and Ephraim contained the highest percentage of destroyed Middle Bronze Age sites, followed by the hill country of Manasseh.

TRIBAL DIVISIONS AND TRIBAL BOUNDARIES

Although passages in the Bible describe the allotments of the twelve tribes, the Israelites at this time were composed of separate lineages or clans that developed in time into tribal identities. The clearest indication of this is found in Joshua 17:14–18, in which the "tribe of Joseph" complains that they do not have enough land, and Joshua tells them to clear the forest in the hill country. As archaeologist Lawrence Stager notes: "the number and composition of the tribes fluctuated through time with changes in demography and geography. As fusion and fission occurred among clans, some rose to tribal status."[34] The process of fission was dominant among the northern tribes, while fusion characterized the southern Israelites.

The oldest tribal divisions are reflected in boundaries that geomorphological evidence dates to the earliest part of the Late Bronze Age or even earlier, when the Dead Sea extended both farther north and farther south than it did later on. From about 2140 to 1445 (or 1500) B.C.E. the Dead Sea was in the process of falling from an earlier highstand that existed *prior to the fifteenth century* B.C.E. to below 380 mbsl (meters below sea level).[35] The northern and southern tongues or bays could have existed only when the water level was above 370–375 mbsl (see figure 9.3). Sometime between

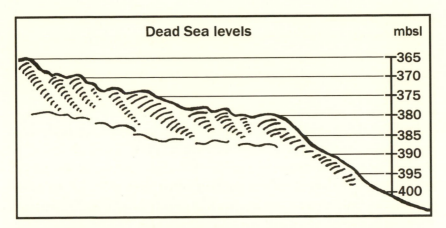

FIGURE 9.3. Schematic sketch of Dead Sea levels. mbsl = meters below sea level. With a higher Dead Sea of about 370–375 mbsl, northern and southern bays existed that are reflected in the earliest Israelite tribal boundaries (see text).

1440 and 1120 B.C.E., the Dead Sea reached the extreme low level of 410.5 mbsl; lake levels remained low between about 1000 and 550 B.C.E.[36]

The southern boundary of Canaan (Numbers 34:3) and of Judah (Joshua 15:2) refers to a southern tongue of the Dead Sea that could have existed only during this period of higher (380–370 mbsl) sea level, in or before the early Late Bronze Age.[37] In a similar fashion, Joshua 15:5–6 and 18:19 describe the Judah-Benjamin border from the northern bay of the Dead Sea at the mouth of the Jordan River to the slope or shoulder of Beth Hoglah and then westward. Beth Hoglah, near Ein Hajla or Deir Hajla, is about five to six kilometers north of the present mouth of the Jordan, so running the border from the river mouth to this area would require a meaningless loop north. But with the Dead Sea at about 375 mbsl, the border would start at the top of the northern bay, go due west to Beth Hoglah, and westward from there.[38] Even the boundary between Reuben and Gad on the east side of the Jordan River indicates a northern bay and high Dead Sea level, because all the cities of the Jordan Valley were given to Gad (Joshua 13:27) except for Beth Jeshimoth (Joshua 13:20). With a high early Late Bronze Age (pre-fifteenth century B.C.E.) Dead Sea level, Beth Jeshimoth is not in the river valley (Gad's territory) but instead is on the east side of the Dead Sea itself (in the territory of Reuben) (see figure 9.2).[39]

Such early, geomorphologically determined dates for these boundaries functionally preclude an Israelite conquest of Canaan in the thirteenth century B.C.E., that is, during or after the reign of Ramesses II. They also eliminate a conquest at the end of the fifteenth century B.C.E., following an Exodus date derived from 1 Kings 6:1 (see chapter 1).

ISRAELITE SETTLEMENT IN THE CENTRAL AND NORTHERN HILL COUNTRY

Many of the early traditions in Joshua 1–9 were probably part of a combined tradition of the tribes of Joseph (Ephraim and Manasseh and also Benjamin), before Benjamin split off from the original group—Benjamin means "son of the south" and his position as last born of Rachel indicates this tribe's later emergence from a larger Israelite group.[40] Also separating from Ephraim and Manasseh were Gilead (another name for Gad) and several clans of Manasseh, who settled west of the Jordan (see figure 9.1).[41]

There are hints that other tribes also originated in the Ephraim-Manasseh heartland. Beriah, the son and founder of the northern tribe of Asher (see figure 9.1), is listed as a son of Ephraim in 1 Chronicles 7:23. His sister is listed as a daughter of Ephraim, while his grandson Birzaith lends his name to the Iron Age settlement of Bir ez-Zeit (Khirbet Bir Zeit), fifteen miles north of Jerusalem; also, a section named Asher is mentioned as a part of Manasseh in Joshua 17:7.[42] Even more interesting is the thirteenth century B.C.E. Egyptian papyrus Anastasi I that mentions some "shasu" (nomads or bedouin) living in the tall bush near a pass in the central hill country of Manasseh: "Their hearts are not mild, and they do not listen to wheedling." The name of these shasu is Asher.[43]

The sons of Issachar lived in the hill country of Ephraim and Manasseh, and the archaeological evidence points to a Manassehite origin for the settlers of the territory of Issachar.[44] The boundary lists in Joshua 15–19 indicate that the early tribal confederacy included only Ephraim, Manasseh, Benjamin, Zebulun, Asher, and Naphtali. By the time of Deborah (Judges 4 and 5) it had expanded to include Issachar, Dan, Reuben, and Gilead (Gad) as well.[45]

The fact that many Canaanite cities are on the periphery of the tribal territories is a clear indication that the nomadic Israelites settled around and between the cities, which served them as trading centers; they did not conquer these cities, at least not at the time the boundaries were formed.[46] In Galilee, the tribal boundaries for the most part followed valleys, again indicating that the Israelites themselves lived in the sparsely populated mountainous regions. In the fourteenth century B.C.E. the Galilean region was said to be inhabited by 'Apiru.[47]

The settlement of the southern Israelites, known biblically as the tribes of Judah and Simeon and part of the tribe of Levi, is much less straightforward than that of the northern Israelites. Certain biblical traditions point to an Israelite invasion of Canaan from the south by a group that split off from the main body somewhere in the Wadi Arabah. Judges 1:16–17 says the descendants of Hobab the Kenite went up with the people of Judah from the city of palms ('Ain Hosb, in the Wadi al-'Arabah not far from Petra) into the Negev near Arad and tells how Judah and Simeon took and destroyed Hormah, on the border between the Judean hill country and the Negev (see figure 9.1). In Numbers 14:39–45, the conquest of Hormah was a prelude to the unsuccessful invasion attempt under Moses.[48]

Levite towns were in the south, evidence that most of the Levites originally settled in southern Judah along with the Simeonites, who settled near Beer-sheba. Some Levites, however, remained in Ephraim, among them the ancestors of the Mosaic priesthood later found at Shiloh and Dan.[49] The Levites also probably became the repositories of the Israelites' traditional history when their responsibility as guardians of the ark evolved from a military to a ritualistic role, as religious practices developed around the ark.

The conquest of Hebron, the main city in the southern hill country, is credited to both Joshua and to Caleb, one of the two faithful spies from the first reconnaissance under Moses (see chapter 6).[50] Debir, not far from Hebron, was conquered by Othniel, Caleb's nephew.[51] In fact, Hebron and Debir were probably both conquered by Kenites or Kenizzites and the story of this conquest later integrated into the overall southern Israelite oral tradition. The only archaeological evidence for the conquest of Hebron is much later, at the end of Late Bronze Age II just before the onset of the Iron Age. However, most of the ancient city is still unexcavated.[52]

The northern Judean hill country was settled by the Ephrathites. The name means "men of Ephraim," and this group, which includes the ancestors of King David, clearly came from Ephraim. In 1 Chronicles 2:24 it is recorded that Caleb married Ephrathah who bore him a son named Ashhur, the father of Tekoa, actually Khirbet Teku, five miles south of Bethlehem. This genealogy reflects the peaceful coming-together of the Ephrathites and the Calebites.[53]

Othniel, son of Kenaz, is Israel's first judge, who delivers his people from the tyranny of King Cushan-rishathaim of Aram-naharaim.[54] The name Aram in Aram-naharaim was probably originally Edom,[55] and the king's name, "Cushan of double wickedness," reflects the ancient name of for Edom, Cushu. Thus the story of Othniel came from a group of Cushites who fought a Cushan/Edomite tribal leader and then merged with the southern Israelites.

The tangled genealogies and varied stories in the biblical accounts point to close relationships among the pastoral nomads of the Judean hill country, those of Edom or Seir (as reflected in the Othniel stories), as well as the Jerahmeelites of the Negev. In a study of biblical family names, Avi Ofer found that Judah shared 35% of its names with the people of Edom and 33% with the tribe of Simeon. Only the Israelite tribe of Reuben had a higher percentage of shared names with Judah—37%.[56] In fact, scholars have long noted that the tribes Reuben and Judah share two clan names, Hesron and Karmi. In the Bible Reuben is the firstborn of Jacob, and its tribal border, as discussed earlier in this chapter, existed in the earliest part of the Late Bronze Age.

Reuben's territory was east of the Jordan in what later became the northern part of Moab.[57] However, Frank Moore Cross noted that: "place names associated with Reuben are found on the west bank of the Jordan along the northern boundary of Judah. . . . It may be noted that all of these sites follow the main ancient road from the ford [across the Jordan] immediately north of the Dead Sea up the Wâdī Dabr by the stone of Bohan [son of Reuben], modern Hajar el-'Asba', through the 'ēmeq ākōr [Valley of Trouble], the modern el-Buqê'ah, then north to Jerusalem, or south to Hebron" (see figure 9.1).[58] Cross suggested that a western division or offshoot of Reuben penetrated from Shittim to Gilgal on the Jordan and along this ancient route through the Valley of Trouble into the territory that later became known as "Judah."[59]

Judah was probably originally a geographic name referring to the hill country from north of Bethlehem to south of Hebron. As often happens in oral traditions, it became anthropomorphized into a person. Eventually, all these disparate population elements—western Reubenites, Simeonites, Calebites, Kenizzites, Edomites, Ephrathites, and Jerahmeelites—came peacefully together under the tribal rubric Judah, possibly during the reign of King David in the tenth century B.C.E.[60]

POPULATIONS AND MALARIA IN LATE BRONZE AGE CANAAN

Archaeological evidence of a series of isolated cultic shrines or sanctuaries (unrelated to any settlements or beyond the boundaries of towns) and of cemeteries not adjacent to any permanent sites suggests a substantial population of pastoral nomads living in the central hill country of Canaan in the Late Bronze Age.[61] They would have lived principally in the desert fringe (the desert fringe of Manasseh alone could have supported 1,500 to 2,000 nomads[62]) and in the eastern part of the hill country, trading with the peoples of the towns (Shechem, Bethel, etc.), burying their dead in cemeteries located away from cities, and maintaining their own religious cult centers at Shiloh, and near (but not *in*) Shechem and Lachish.[63] Na'aman gives a rough estimate of seven to ten thousand for the overall number of nomads in Canaan in the Late Bronze Age,[64] a number not out of line with the expected increase of the Israelites.

Archaeologists and ethnographers have recently come to realize that past population fluctuations were much greater than previously suspected— certain ethnic groups have expanded in both numbers and territory while others have gone extinct.[65] In South America, for example, the Yanomami, who live in the Amazonian rainforest, have expanded by two or three times in the past hundred years.[66] The survival of young children is the most important factor in determining population increase, and thus societies in which these children are more likely to survive will "outcompete" their neighbors in terms of population growth; climate factors, especially those that affect the survival of children, are also very important.[67]

In premodern times, the lowland areas and interior valleys of Canaan contained many interior drainage networks, resulting in more standing water and swamps. These swamps were breeding grounds for malaria mosquitoes. Malaria takes a high death toll on young children, particularly in areas of higher population density, such as towns and cities.[68] Town dwellers in the lowlands and valleys of Canaan in the Late Bronze Age, with their higher population densities, would be subject to endemic malaria, particularly if they slept outside on rooftops in hot weather, as people in that part of the world have long done (most malaria mosquitoes bite at night). Pastoralists, in contrast, have lower population densities and thus lower incidences of malarial infection. They are surrounded by animals (who are more likely to be targets for mosquitoes) and sleep in tents. The Israelite pastoralists of the Late Bronze Age, living in the hills and highland areas away from the swamps and standing water, with their animals, their lower population densities, and their tents, would have been far less affected by this scourge and thus have fewer deaths of their infants and young children. Consequently, the Israelites would have experienced a higher population growth rate than their town-dwelling and valley and lowland Canaanite neighbors.[69]

Late Bronze Age texts contain many reports of Habiru or ʿApiru in Canaan and elsewhere. As noted before, the term probably included the Israelites but was not limited to them. During this period the term ʿApiru developed a more negative meaning than it had had in the Middle Bronze Age (see chapter 5). It now referred to bands of uprooted people who came down from the highlands into the lowland areas of Canaan and caused trouble for some local rulers and acted as mercenaries for others.[70] Jephthah the Gileadite (Judges 11) was a typical ʿApiru of the Late Bronze Age. The son of a prostitute, Jephthah was forced by his legitimate brothers to flee Gilead. He became the head of an outlaw band, but when Ammonites attacked from the area of today's Ammon, Jordan, the people of Gilead asked Jephthah and his band to defend them. After he beat the Ammonites Jephthah fought with the Ephraimites (Judges 12), although they were, like himself, fellow Israelites.

The most interesting reference to the ʿApiru appears in the early fourteenth century B.C.E. Amarna correspondence from Egypt. These clay tablets, found in the royal city of the pharaoh Akhenaten, include many letters sent by Canaanite rulers to the Egyptian court. These rulers often complain to their Egyptian overlord about raiding Habiru and plead for contingents of Egyptian archers to help defend against these marauders.[71] In one of them, the ruler of Shechem is accused of being in league with the ʿApiru. Another letter from another Canaanite leader accuses the ruler of Sechem of being an ʿApiru himself. Several generations later the ʿApiru were still active around Shechem, much as the Israelites are reported to have been during the time of Abimelech (Judges 9).[72]

Another term that probably includes the early Israelites is the Egyptian word "shasu," which in the New Kingdom period referred to nomadic peoples, usually (but not always) living to the south of the Dead Sea. By the end of the thirteenth century B.C.E. "shasu" are said to be from Edom, but an Egyptian list that reflects an earlier, fifteenth century topographical list names six groups of shasu including the Shasu of Seir, the Shasu of *Rbn*, the Shasu of Sam'ath (probably a clan of Kenites), the Shasu of *Wrbr* (probably near the Wadi el-Hesa), and the Shasu of *Yhw*.[73] In this context *Yhw* is the name of a place, but most scholars agree it is an early form of the name of the Israelite god Yahweh.[74] And although some scholars simply equate the shasu with the Edomites, the different names in the Egyptian list point to a more complicated situation, in which some of the groups of shasu were connected to occupants of what eventually became the tribal territory of Judah.

The Judean hill country during the Late Bronze Age was sparsely inhabited, the only substantial settlement being Khirbet Rabud, a site of only two hectares.[75] Almost all of the population in the territory that became Judah was nomadic and would qualify for the Egyptian term shasu. The western offshoot of the Reubenites may have been the Shasu of *Rbn*. The Kenites would have been represented by the Shasu of Sam'ath, the Moabites by the Shasu of *wrbr*, the Edomites by the Shasu of Seir, and the southern Israelites (Simeonites, Levites, possibly Jerahmeelites) by the Shasu of *Yhw*.

The very real probability that some of the people the fifteenth and fourteenth century B.C.E. labeled "shasu" by the Egyptians were in fact people who would become known as Israelites is important because, during those centuries, Egyptian texts contain reports of large numbers of shasu taken as captives or slaves to Egypt. Some of these, most notably the Shasu of *Yhw* and the Shasu of *Rbn* would have found themselves in the land their ancestors had left so precipitously about 175 years before. This time around, just as described in the book of Exodus, the Israelites actually were slaves.

CHAPTER TEN

Back to Egypt

CAMPAIGNS OF THE WARRIOR PHARAOHS

While the Israelites and other peoples were invading Canaan—that is, in the second half of the sixteenth century B.C.E.—Egyptian kings contented themselves with the taking of the Canaanite cities of Sharuhen and Gaza. Rather than referring to campaigns in Canaan, Egyptian records for this period contain accounts of expeditions and raids farther north, into Lebanon and Syria.[1]

Ahmose, the first ruler of Egypt's Eighteenth Dynasty, reigned for twenty-five years and was succeeded by his son Amenophis I. Amenophis I's successor was Tuthmosis (or Thutmose) I who had married Amenophis I's sister. Tuthmosis I was a commoner of obscure origin, but he was the first of the Eighteenth Dynasty's warrior kings, leading campaigns into Nubia and to the northern Levant as far as the upper Euphrates River, where he erected a stela to commemorate the event. His son, Tuthmosis II, was married to his half-sister Hatshepsut, Tuthmosis I's daughter. Upon the death of Tuthmosis II, Hatshepsut became regent for her stepson, Tuthmosis III, who was a young child at the time. Not long after assuming the regency, however, Hatshepsut took over power and became queen in her own right. It was only twenty-two years later that Tuthmosis III became the actual ruler of Egypt upon Hatshepsut's death. His reign, counted from his childhood accession, lasted for nearly fifty-four years.[2]

Tuthmosis III was the greatest warrior pharaoh of the Eighteenth Dynasty, and he was the first to campaign actively in Canaan. His father, Tuthmosis II, had led a raid into the Sinai and Negev and returned with a number of shasu prisoners, and Tuthmosis I's army had marched through Canaan on its way to Syria, but Tuthmosis III led an army through Canaan to put down a revolt of the Syrian princes led by the ruler of Kadesh, an important city-state on the northern Orontes River. These princes and their armies moved south into northern Canaan just as Hatshepsut was dying.[3] After only a month as sole ruler of Egypt, Tuthmosis III set out with his foot soldiers and chariots along the coastal road across the Sinai and north through Canaan until he came to a halt just south of the Carmel Ridge. North of the Ridge were the armies of the Syrian princes, just outside the town of Megiddo.

There were three routes open to the Egyptians: a detour to the north, another to the south, or a narrow, steep route through the Aruna Pass directly across the Ridge. Against the advice of all his generals, Tuthmosis III decided upon the direct route, up a steep and difficult path where the chariots would at times have to be lifted manually. He and his chariot were first in line. At dawn the next day a surprised Syrian army beheld the entire Egyptian army on their flank. Before the Syrians could redeploy to face their opponents head-on, the Egyptians attacked down the hill. The forces of the Syrian princes broke and fled, chased by the victorious Egyptians. A few of their princes made it into Megiddo, but it did them little good in the long run—Tuthmosis III kept the town under siege until it surrendered seven months later.[4]

Once he had captured the city, Tuthmosis extracted loyalty oaths from the enemy princes and sent most of them back to their respective towns. When Tuthmosis returned to Egypt he had a record of his victory and his subsequent campaigns carved in stone in the temple at Karnak, providing a detailed history of his military campaigns. Later in his reign, he had a tall black granite tablet engraved with the words of the god Amon-Re. Inscriptions of previous monarchs had the ruler address the god; here Amon-Re himself speaks of the exploits of Tuthmosis.[5]

Tuthmosis III returned to Syria and Canaan often, exacting tribute and loyalty oaths. Later he installed commissars in Canaanite cities to keep the tribute flowing back to Egypt. Much of this tribute was in the form of slaves: in year 30 of Tuthmosis III's reign, thirty-six men, 181 male and female servants; in year 31, 492 prisoners of war; in year 33, 579 male and female servants with their children; in year 34, 602 male and female servants; in year 38, fifty prisoners of war and 522 male and female servants.[6] When large numbers of prisoners or slaves were acquired, the entire group was handed over as a unit to one of the great temples to work on the temple estates.[7] The tomb of Rekhmire, Tuthmosis III's vizier for the south, contains paintings of prisoners of war, some of them from Syria-Canaan, making bricks for a ramp that was used in the construction of a temple.

Also in the picture are stick-wielding Egyptian overseers.[8] Egyptians, by the way, habitually used straw as chaff in their bricks.

Tuthmosis III's last recorded campaign in Syria-Canaan was in year 42 of his reign, but as late as year 50 (when he was fifty-five to sixty years of age) the pharaoh was still campaigning in Nubia.[9] For the last two years of his life Tuthmosis III's son, Amenophis II, was co-ruler. This young man was eighteen years old on his accession. In the last year of his father's life, Amenophis II led his own campaign to Syria-Canaan. By the time he returned to Egypt—or very soon after—his father, Tuthmosis III, was dead, on the thirtieth day of month seven in the fifty-fourth year of his reign, by modern calculation (using the higher chronology, see table 7.1), on the 17th of March, 1450 B.C.E.[10]

THUTMOSIS III'S NAVAL BASE

Tuthmosis III soon realized that it would be easier for the greater part of his army to sail to Syria than to make the long and arduous trek through Canaan (the chariots and horses probably still traveled overland). Around the year 30 of his reign he constructed a large naval base where a fleet was built that could transport most of his army to the northern Levant coasts; there they would disembark and march east to confront the Syrian city-states.[11] These ships could also carry the pharaoh's tribute and slaves back to Egypt. That at least some of these slaves were Israelites is indicated by this warning at the end of Deuteronomy 28: "The Lord will bring you back in ships to Egypt, by a route I promised you would never see again; and there you shall offer yourselves for sale to your enemies as male and female slaves, but there will be no buyer."[12]

Manfred Bietak has recently uncovered an extensive early Eighteenth Dynasty occupation at Tell el-Dab'a (specifically, at 'Ezbet Helmi) that included a great palace and storage facilities, probably for a temple, a military base, workshops, and probably a dockyard. Within the settlement he found scarabs of pharaohs from Ahmose to Amenophis II.[13] Reactivating the old Hyksos capital of Avaris (probably never completely abandoned) makes a great deal of strategic sense, given the site's position near the head of the Pelusiac Branch of the Nile and as the terminus of the northern land route through the Sinai to Asia (the Way of Horus, see figures 2.1 and 10.1). The fortress at Tell Hebua I (known as Tjaru), on a peninsula between the open sea and a brackish coastal lagoon (the Shi-Hor), was also reactivated in early New Kingdom times and controlled

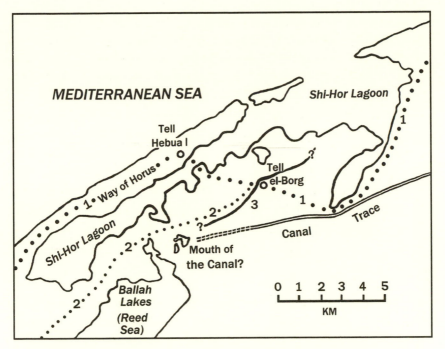

FIGURE 10.1. The northeastern edge of the Nile Delta and northwestern Sinai reconstructed from: Sneh, Weissbrod, and Perath; Marcolongo; Valbelle et al.; Bietak (*Avaris*); and Hoffmeier (*Israel in Sinai*). 1 = the Way of Horus, the main route between the northern Delta and the Sinai in Second Intermediate Period and New Kingdom times; 2 = proposed secondary route that joined the Way of Horus near Tell el-Borg (see text); 3 = channel found during excavations at Tell el-Borg (see Hoffmeier, *Israel in Sinai*). Canal trace from Sneh, Weissbrod and Perath and Hoffmeier (*Israel in Sinai*).

the main land route from the forward base at Avaris to the Sinai and Canaan.[14]

After becoming sole ruler Amenophis II did not long continue the use of Avaris (by whatever name it was then known). Instead, he built another dockyard and naval base called Peru-nefer (Happy Journey) near Memphis.[15] This base was farther from the Egyptian empire in Syria-Canaan and farther from the overseas timber sources needed to build the Egyptian ships. Why did Amenophis II relocate his forward base back from the Delta proper to Memphis?

One factor was undoubtedly the control of his labor force. The temple and palace complexes required large numbers of slaves to carry out various tasks and to work in the fields, and the dockyard too may have utilized a great many unfree workers. If these people were originally from Syria and Canaan—and Amenophis II brought thousands of prisoners from these lands to Egypt—they would be tempted to escape eastward from the Delta through the shallow reedy lakes that covered so much of the isthmus that separated Egypt from the Sinai Peninsula. In the Sinai proper the Eighteenth Dynasty Egyptian forts were small waystations manned by local tribesmen. Only in the thirteenth century B.C.E. did the Nineteenth Dynasty exercise sufficient control of the northern Sinai Peninsula (by means of a crocodile-infested border canal, a fortress in the Wadi Tumilat, and a line of massive roadside fortresses garrisoned by Egyptian troops along the northern land route from the Delta through the Sinai itself) to build another great urban center in the eastern Delta—Pi-Ramesses, just north of Tell el-Dab'a.[16]

THE PHARAOH OF THE EXODUS

In 1963 J. G. Bennett suggested that Tuthmosis III was the pharaoh of the Exodus and that the eruption of Santorini was connected to both the destruction of Atlantis and the Exodus from Egypt.[17] Basing his calculations on the statement in 1 Kings 6:1 that the Exodus occurred 480 years before the beginning of Solomon's temple, Bennett put the Exodus in 1447 B.C.E., which he thought was the year of Tuthmosis III's death. In 1982, using the genealogical information available for the Egyptian royal family, William Shea argued that the Exodus occurred during the co-rulership of Tuthmosis III and Amenophis II, when Amenophis was on his first campaign in Asia. The pharaoh drowned in the Exodus, he concluded, was Tuthmosis III. In revenge his son, Amenophis II, conducted a brutal campaign in Canaan and harbored a life-long hatred of Semites, both of which are documented in Egyptian records.[18]

In chapter 4 we looked at the two exodus stories found in the Hebrew scriptures. In the more recent of these, the exodus-expulsion, the Israelites are slaves who ask permission of Pharaoh to go on a three days' journey into the wilderness to make a slaughter-offering to their god. But Pharaoh refuses their request and makes them work harder, denying them the straw for their bricks, so that the Israelite supervisors are beaten by their Egyptian taskmasters when they fail to make their daily quota of bricks. The Israelite supervisors complain to Pharaoh, but to no avail. Pharaoh comes across as extremely arrogant and breaks off negotiations. After that the Israelites kill and eat the sacrificial lambs and bread (cf. Genesis 31:54). That evening the firstborn of Egypt die. Pharaoh connects the deaths with his Israelite slaves and expels them: "Rise up, go away from my people, . . . Go worship the Lord as you said . . . and be gone" (Exodus 12:31, 32). Exodus 11:1 also refers to Pharaoh driving the Israelites away.

THE EXODUS IN EGYPTIAN TEXTS

Several Egyptian sources may relate to this expulsion. The first is in the work of third century B.C.E. Egyptian historian Manetho. In quoting Manetho, the Jewish historian Josephus substitutes the names Tethmosis and Thummosis in the story of the expulsion of the Shepherds or Hyksos from Egypt, whom Josephus equates with the Jews. For example Josephus writes: "Tethmosis, the king who drove them [the Shepherds] out of Egypt . . ." and "Moses was the leader of the Jews, as I have already said, when they had been expelled from Egypt by King Pharaoh whose name was Tethmosis." The confusion went the other way in another ancient writer's version of Manetho: Syncellius, following Africanus, reports that Moses went forth from Egypt in the reign of A[h]mose.[19] This confusion seems to stem from the mixing of two expulsion traditions, one relating to the expulsion of the Asiatic Hyksos to southern Canaan by Ahmose, and the other involving a pharaoh named Tuthmosis, who also expelled a group of Asiatics, in this case southern Israelites, to Canaan.

In the account by Artapanus mentioned in chapter 5, "Pharethothes" was king of Egypt in Abraham's time. This name is a conflation of "pharaoh" and "Thoth," the latter being the root in the name Tuthmosis. As psychologists studying serial recall discovered, events from the end of a story often become fused with ones from the beginning.[20] This characteristic of memory probably explains why the name of the pharaoh in the last part of the story of the Israelites in Egypt, Pharaoh Tuthmosis, found its way into the story of the first trip to Egypt, the trip made by the Israelites' patriarch, Abraham.[21]

The third Egyptian source is the inscription on the naos originally found at El-Arish, mentioned in chapter 1. The inscription was carved on black granite, and in the nineteenth century when archaeologists first discovered it, it was being used as a water trough. The text was destroyed on one side, and there was a good deal of damage elsewhere. Hans Goedicke believes that the naos was made just before the Persian invasion of 525 B.C.E.[22] The inscription gives a folkloric version of Egyptian history, including references to the Hyksos (called the companions or children of Apophis and evildoers of the desert) into the early Eighteenth Dynasty. Instead of referring to actual kings the inscription follows the typical practice in ancient

Egypt and uses the names of gods, in this case the gods Shu, his twin sister and wife Tefnut, their son Geb the earth god, and Ra-Harakhte, the god of the eastern horizon. The English translation of the text also refers to another god, Thum or Tum.[23]

The inscription describes the building of a palace-temple complex on the easternmost frontier of Egypt above Memphis, actually two temple enclosures joined by an avenue, which is the common New Kingdom form.[24] The text also refers to a great storehouse in front of the temple. One of the bodies of water east of this complex was called the Place of the Whirlpool. The king (the term pharaoh is not used) fortified the mounds or hills that guarded the roads leading into Egypt from the east to protect the Delta from the Asiatics, called the children of Apophis. He is also said to have conquered the whole earth and to have been always at the head of his troops. But "[sickness came upon him?] confusion seized the eyes[?] and evil fell upon the land and there was a great upheaval in the palace."[25] The rebels carried disorder to the household of the king himself. His majesty King Shu "departed to heaven" [he died] with his attendants. Another part of the text says that King Shu had died and no one left the palace during the space of nine days, during which time there was such a tempest and darkness that neither men nor gods could see the faces of those next to them. The text then says the king's son Prince Geb was wandering around looking for his mother and after King Shu's death his son Prince (now King) Geb retrieved his mother at Pi-Kharoti, where she had gone to see what had happened to King Shu. The text also says his majesty Ra-Harakhte fought with the evildoers (or rebels) at the Place of the Whirlpool.

The naos inscription presents a very confused account, written many centuries after the events behind it had passed into folklore, but it does seem to relate to a real historical incident. A conquering ruler built or maintained a great palace-temple complex on the eastern edge of his domain above Memphis (the complex seems to have been started by an earlier king). There was some sort of upheaval or rebellion by Asiatics in the palace complex, and a malady, possibly a sickness. The king died with his attendants, possibly while fighting the rebels at the Place of the Whirlpool east of the palace-temple complex, near a place called Pi-Kharoti. There was period of storm and intense blackness (the figure nine days can probably be discounted because nine is an Egyptian ritual number). His son becomes king and discovers that "certain Asiatics carried his [the king's] scepter, called Degai, who live on what the gods abominate."[26] It's the Asiatics who live on what the gods abominate.

The idea that Asiatics are connected with abominations, found in the El Arish inscription, goes back at least to the reign of Hatshepsut who claimed to have driven off Asiatics, called "the abomination of the great god [Re]."[27] During the New Kingdom, Egyptians revered sheep as being the residence of the soul or *ba* of the god Amon-Re (Amon, a Theban deity, was combined with the sun god Ra or Re in this period), and a cult revering the wild sheep existed in Deir el Medinah, west of Thebes, during the Nineteenth Dynasty. Even earlier, during the reign of Amenophis III of the Eighteenth Dynasty, there was a proliferation of images of Amon as a sheep in Thebes.[28] The idea, mentioned in the plague accounts in Exodus 8:26, that Israelite sacrifices of sheep would be seen as offensive or an abomination to the Egyptians and their god Amon-Re fits with this time period.

There is another Egyptian tale that relates to the slaughtering of animals sacred to the Egyptians, found in the work of Manetho (as quoted by Josephus). Manetho repeats a fragment of legendary Egyptian history in which king Amenophis, on the advice of his seer, also named Amenophis, cast all the lepers and other polluted persons into the stone quarries east of the Nile and later let them live in the deserted city of Avaris. There they revolted under a priest of Heliopolis called Osarseph and sacrificed and butchered animals sacred to the Egyptians. They also allied themselves with the Shepherds (that is, the Hyksos) who had previously lived at Avaris. After a campaign in Ethiopia, Amenophis returned to battle the lepers and polluted persons and their allies the Shepherds, defeated them, and pursued them to Syria.[29]

In fact, the first Eighteenth Dynasty pharaoh, Ahmose, *did* open up stone quarries that used Asiatic workers in the twenty-second year of his reign. He also reoccupied and rebuilt part of Avaris, as shown by Manfred Bietak's archaeological excavations at Tell el-Dab'a, after first expelling the Hyksos, some of whom were probably carriers of (or polluted with) the bubonic plague. And though it was Amenophis III who promoted the divine images of Amon as a sheep (and had an advisor also named Amenophis), it was Amenophis II (son of Tuthmosis III) who defeated the Asiatics and campaigned in Syria and Canaan.[30] All of this shows that the Egyptians could confuse, conflate, or syncretize their legendary history as easily as anyone else. But this story does contain enough Eighteenth Dynasty connections to suggest that sometime during the earlier part of this dynasty Asiatics were living in the vicinity of Avaris, revolted, and sacrificed animals that were sacred to the Egyptians, much like the revolt mentioned in the El Arish text. The revolt between the "lepers" and the gods of Egypt, centering on illegal sacrifices, is much like the confrontation between the Israelites and Pharaoh recounted in Exodus 5–10.[31] In this account, Moses and Aaron repeatedly ask Pharaoh for permission to make a three days' journey into the wilderness to celebrate a festival to their god. Pharaoh—rightly from an Egyptian point of view—claims not to know or to heed the god of the Israel and eventually (Exodus 10:28) expels Moses from his presence.

COMPARING THE EL ARISH INSCRIPTION WITH THE BIBLICAL TEXT

The El Arish naos account contains marked similarities with the biblical exodus-expulsion story. In the Egyptian text, the king goes to his palace-temple-storehouse complex in the Eastern Delta, while his son, Prince Geb, is off on a journey. Disorder by rebels is carried into the king's household and there is, apparently, sickness. In the biblical story the Israelites ask Pharaoh for permission to hold their slaughter offering in the wilderness (Exodus 3:18, 5:1b). This can only have been their annual covenant-renewal sacrifice. Since the Israelite lunar calendar falls behind the Egyptian solar calendar unless adjustments are made every few years, the Israelites may have been making their request to Pharaoh a little *before* the spring equinox. When refused permission to leave, they kill the ritual lambs anyway, right where they live. The animals must have been taken without permission from the palace and temple flocks, the act of rebellion referred to in the El Arish text. Isolated within their own dwellings and eating their sacrificial lambs and covenant meal, the Israelite slaves do not fall prey to the illness that strikes the Egyptians. This malady seems to have preferentially killed children, whose immune systems are far less developed than those of adolescents or adults. Food poisoning, such as an outbreak of *Salmonella* or *Escherichia coli*, often has this sort of infection and mortality pattern.[32] The tainted food, prepared in the central kitchen of the palace-temple complex, would have been passed to all but those children and infants who were still breast-feeding (not the "firstborn").

In the biblical account (Exodus 11:2,12:35–36) the Israelites leave following Pharaoh's expulsion order, taking some gold, silver, and clothing, possibly offerings from desperate parents hoping for divine intervention to save the lives of their sick children, as implied in Exodus 12:32b. Then Pharaoh changes his mind (Exodus 14:5b). Deciding that he doesn't want to lose his slaves, he calls out his chariots and sets off in hot pursuit. The

El Arish text implies something quite different: that the Asiatics had the royal scepter. Retrieval of his scepter and his slaves are both good reasons for Pharaoh to chase after the Israelites.

THE ROUTES OF THE ISRAELITES AND OF PHAROAH'S ARMY

From Avaris both the Israelites and the pursuing Egyptians would have moved east along the Way of Horus until they reached Tell Defenneh (see figure 2.1). The main route, the Way of Horus, proceeded northeast up the narrow peninsula to the fortress of Tjaru (Tell Hebua I—see figure 2.1). This peninsula is actually a kurkur sandstone ridge, a structural ridge that is part of a system of faults and lineaments which extends well beyond the Delta to the northeast and southwest.[33] The stabilized sand dunes covering this ridge are the fortified mounds or hills mentioned in the El Arish text. Southeast of this uplifted sandstone ridge is a tectonic trough once filled with a brackish coastal lagoon, the Shi Hor.[34] This lagoon (or at least its northern or northeastern segment) was open to the Mediterranean Sea through several breaks in the kurkur ridge. A peninsula jutted out from the shore opposite Tjaru. The road went from Tjaru across either a bridge or dike to this peninsula and southeast from there. If the road was atop a dike, then the southern (or southwestern) part of the lagoon was cut off from the sea.[35]

There was another land route, one that went east from Tell Defennah across the isthmus of Qantara between the Shi-Hor and the northern edge of the Ballah Lakes.[36] It would have joined the main Way of Horus near Tell el-Borg, upon which was a fort built by Tuthmosis III, on the east side of a water channel that most likely flowed into the northeastern Shi-Hor (see figure 10.1).[37] For the Israelites, this route would have been an easier way home, since it bypassed Pharaoh's largest fortress and may even have offered a way around the smaller fort at el-Borg. This secondary route was a bit longer and would have crossed low-lying ground in places and thus would not have been as suitable for horses and chariots. Tuthmosis and his chariots would have used the shorter main route, from Tell Defenneh northeast across the kurkur sand ridge to Tjaru. At Tjaru he could have picked up more chariots and men if he needed them.

ENCOUNTER AT THE EDGE OF THE MEDITERRANEAN SEA

There were probably no more than several hundred Israelite slaves in the group expelled by Pharaoh, principally nomadic shasu from the southern

Israelite tribes. In Exodus 14:2 the Israelites are camped in front of Pi-hahiroth, between Migdol and the sea, in front of Baal-zephon. Pi-hahiroth is remarkably similar to the name Pi-Kharoti or P^3-h^3-r-ty found in the El Arish naos text. Both terms refer to the mouth of a canal. A recent interpretation of Baal-zephon by James K. Hoffmeier associates this name with the "waters of Baal" somewhere near the northern edge of the Ballah Lakes, the biblical Reed Sea.[38] North of the Ballah Lakes was the fort at el-Borg. Hoffmeier proposes that the biblical Migdol (which means a tower in Hebrew) was located between the southeastern tip of the coastal lagoon and the Ballah Lakes, but el-Borg in Arabic means the tower, and it is a good candidate for the biblical Migdol.[39]

According to Exodus 14:9–10, as the Israelites camped by the mouth of the canal (Pi-hahiroth), in front of Baal-zephon, they saw Egyptian soldiers on their fast-moving chariots, coming up behind them (see table 10.1). They realized that Pharaoh had changed his mind, and now they were trapped. If they were on the secondary road, northeast of them was the fort of el-Borg, and beyond it the canal itself. North and west of them was the coastal lagoon, a branch of the Mediterranean Sea, and south of them the "waters of Baal," the Ballah Lakes or Reed Sea.

The biblical text implies that it was at the end of the day. Pharaoh, reaching his fortress of Tjaru, also stopped to rest his horses and organize his forces for an attack the following morning. Pharaoh and the Egyptians also knew that the Israelites were trapped.

Exodus 14 does not give us a clear picture of what happened when the Egyptians caught up with the Israelites. As one scholar has said: "the story as it now stands is a composite of several traditions which, having been brought together, *fail to present a clear picture of a comprehensible event* [his italics]."[40] In one version, as the Egyptian chariots approach, Moses stretches out his hand over the sea and the waters split, allowing the Israelites to walk between the waters. When the Egyptians follow, Moses stretches out his hand again and the waters return, drowning the Egyptians. According to scholars, this version of the Miracle of the Sea, and the names Baal-zephon and Pi-hahiroth, belong to the tradition in the "P" or Priestly source.[41]

A second version of the encounter is also preserved in Exodus 14, one usually attributed by scholars to the "J" source.[42] This second version contains the pillar of cloud (the fire is thought to be a later addition), which is in front of the Israelites until they see the Egyptians pursuing them. Then the pillar moves between the Israelites and the Egyptians and the two

TABLE 10.1.
Miracle of the Sea Texts in Exodus 14 and 15

14:2	Tell the Israelites to . . . camp in front of Pi-hahiroth, between Migdol and the sea, in front of Baal-zephon; you shall camp opposite it, by the sea.
14:5b–7, 8b–10	"What have we done, letting Israel leave our service?" So he [Pharaoh] had his chariot made ready, and took his army with him; he took six hundred picked chariots and all the other chariots of Egypt with officers over all of them. (8b–10): and he pursued the Israelites, who were going out boldly. The Egyptians pursued them, all Pharaoh's horses and chariots, his chariot drivers and his army; they overtook them camped by the sea, by Pa-hahiroth, in front of Baal-zephon. As Pharaoh drew near, the Israelites looked back, and there were the Egyptians advancing on them. In great fear the Israelites cried out to the Lord.
14:17b–18	And so I [God] will gain glory for myself over Pharaoh and all his army, his chariots and his chariot drivers. And the Egyptians shall know that I am the Lord, when I have gained glory for myself over Pharaoh, his chariots, and his chariot drivers.
14:19–20	The angel of God who was going before the Israelite army moved and went behind them, and the pillar of cloud moved from in front of them and took its place behind them. It came between the army of Egypt and the army of Israel. And so the cloud was there with the darkness, and it lit up the night[a]; one [army] did not come near the other all night.
14:24–25	At the morning watch the Lord in the pillar of fire and cloud looked down upon the Egyptian army and threw the Egyptian army into panic. He clogged [or removed] their chariot wheels so that they turned with difficulty. The Egyptians said, "Let us flee from the Israelites, for the Lord is fighting for them against Egypt."
14:21 (part.), 23 15:9	The Lord . . . turned the sea into dry land. (23): The Egyptians pursued, and went into the sea after them [the Israelites], all of Pharaoh's horses, chariots, and chariot drivers. (15:9): The enemy said, "I will pursue, I will overtake, I will divide the spoil, my desire shall have its fill of them. I will draw my sword, my hand shall destroy them."
15:8	At the blast of your [God's] nostrils the waters piled up, the floods stood up in a heap, the deeps congealed in the heart of the sea.
14:27b–28, 15:19a, 15:4–5, 10	As the Egyptians fled before it [the sea], the Lord tossed the Egyptians into the sea. The waters returned and covered the chariots and the chariot drivers, the entire army of Pharaoh that had followed them [the Israelites] into the sea; not one of them [the Egyptians] remained. (15:19a): When the horses of Pharaoh with his

(Continued)

TABLE 10.1. (*Continued*)

	chariots and his chariot drivers went into the sea, the Lord brought back the waters of the sea upon them. (15:4–5): Pharaoh's chariots and his army he [God] cast into the sea; his picked officers were sunk in the Red Sea.[b] The floods covered them; they went down into the depths like a stone. (15:10): You [God] blew with your wind, the sea covered them; they sank like lead in the mighty waters.
14:30	Thus the Lord saved Israel that day from the Egyptians; and Israel saw the Egyptians dead on the seashore.

[a]Martin Noth translates this as the cloud having stayed dark all night (see text).
[b]This could also be translated as Reed Sea.

groups camp, with the darkness between them. That night a strong east wind blows the sea back until there is dry ground, but at dawn the sea returns to its normal depth. Early in the morning the Egyptians panic when they see the cloud and flee into the sea. In this version the Israelites don't move anywhere (and don't follow the cloud) but sit in their camp and watch what happens to the Egyptians. The movement of the waters, although described, doesn't fit into the story.[43]

In different ways, both of these accounts are conflations of the two separate exoduses, the earlier exodus-flight (Moses leading the Israelites, the wind blowing the waters away leaving a corridor of dry land for them to cross the Bitter Lakes or Red Sea) and the later exodus-expulsion story experienced by the Israelite shasu, which features the drowning of the Egyptian soldiers on their chariots that took place "in front of" Baal-zephon by the Reed Sea.

Another version of this latter encounter is contained in two pieces of ancient poetry found in Exodus 15. Both include the lines: "Sing to the Lord, for he has triumphed gloriously; horse and charioteer [or rider] he has thrown into the sea." The verb "to throw" in this song usually means "to shoot" as "to shoot arrows." Some scholars think that this version recounts a battle with the Egyptian forces that ended in the drowning of the enemy.[44] The shorter version of this song (Exodus 15:21), usually called the Song of Miriam, may have been connected to Miriam simply because her name rhymes with the last phrase of the verse.[45] However, as writer Jonathan Kirsch notes:

the most intriguing and important feature of the Song of Miriam is the fact that Miriam did not seem to know—or, at least, chose not to mention—the most colorful and memorable details of the miracle at the sea that so captivated the later authors of the Bible. She said nothing of the parting of the waters, nothing about the crossing of the seabed between walls of water. Neither did she utter the name of Moses or make even an oblique reference to any role he might have played in the miracle at the sea, which raises the provocative notion that he played no role at all because he was not there.[46]

Another part of this ancient poem, Exodus 15:8, describes how the waters behaved: "At the blast of your nostrils the waters piled up, the floods stood up in a heap, the deeps congealed in the heart of the sea." In chapter 3 we saw what happens when a tsunami approaches the shore: the bottom of the wave slows down and drags at the upper part, causing the wave to bunch up to a great height. This closely matches the description in Exodus 15:8.

ERUPTION AND TSUNAMIS

There are textual, traditional, and physical indications that one or more tsunamis occurred in the eastern Mediterranean about this time. Manetho stated that the flood of the mythological Greek hero Deukalion (or Deucalion) occurred in the reign of Mispheagmuthosis and that Mispheagmuthosis reigned for twenty-six years. Misphreagmuthosis is Menkheperre, another name for Tuthmosis III, but twenty-six years was the length of Amenophis II's reign, not Tuthmosis III's.[47] Manetho may have meant that this flood occurred during the co-rulership of Tuthmosis III and Amenophis II, at the end of Tuthmosis III's reign.

Greek archaeologist A. G. Galanopoulos linked Deukalion's flood with the Santorini eruption.[48] In one early version of the Greek myth, the flood comes from the sea, which suggests a tsunami. Estimates of Deukalion's flood vary. The date based on calculations of traditional genealogies is 1430 B.C.E., while a date derived from inscriptions on a pillar known as the Parian Marble (found in the seventeenth century and since destroyed) is 1529 B.C.E. Parian marble dates are usually higher than others, however.[49] Attic Greek folk traditions produce two dates for the flood: 1800 and 1500 B.C.E.[50] Greek myth also includes another, earlier flood, that of Ogyges, who traditionally is the founder of the city of Thebes in Greece.[51] The early Christian writer Julius Africanus put Ogyges' flood in the time of Moses.[52]

At Tel Michal, a coastal site in Israel not far north of Tel Aviv, the sea cliff collapsed twice, once in Middle Bronze IIB and a second time in Late Bronze Age I (1550–1400 B.C.E.). Geologists suggest that tsunamis were involved in these cliff collapses.[53] If Ogyges's flood (as well as the first collapse at Tel Michal) was related to tsunamis from the Minoan eruption of Santorini at the time of the original exodus from Egypt, then other tsunamis may have produced Deukalion's flood and the second cliff collapse at Tel Michal in Late Bronze Age I. Correlating this event with the traditional Greek dates and Manetho's dates makes it most likely to have occurred in the mid-fifteenth century B.C.E.

Tsunamis can be produced by earthquakes, most often if there has been a massive undersea landslide that displaces great masses of water. There is a long record of tsunamis generated by earthquakes in the eastern Mediterranean and particularly in the Aegean Sea.[54] Volcanic eruptions in the ocean can also trigger such landslides if seawater gets into erupting vents, the caldera collapses, or underwater parts of the volcano slump.[55]

Eastward of Santorini on the Aegean Volcanic Arc are the islands of Kos, Nisyros, and Yali, the latter two the remnants of past volcanic eruptions. Between Yali and Nisyros are additional volcanoes on the sea floor. Volcanic rocks from the island of Yali have been dated to the second millennium B.C.E., based on thermoluminescence dating methods, which give an age of 1460 B.C.E. plus or minus 460 years. These rocks were probably produced by an eruption from one of the volcanic calderas now on the sea floor between Nisyros and Yali, sometime in the middle of the second millennium B.C.E.[56]

Yali is a small islet shaped a bit like a bow tie, tucked between Nisyros to the south and Kos to the north (see figure 3.1). Minoan pottery has recently been recovered from volcanic pumice deposits located on the isthmus (the "knot" of the tie) between Yali's northern and southern sections. The most notable ceramic piece from this collection is the top of a painted beaked jug, which the excavators call Late Minoan IA but which bears a striking resemblance to mature Late Minoan IB forms.[57] New radiocarbon dates are consistent with the long-held view that the Late Minoan IB–Late Minoan II transition took place about 1450 B.C.E.[58] The Minoan pottery from Yali is said to have been found within wind-blown pumice. On the southern end of Yali, geologist Jörge Keller found post-Neolithic prehistoric pottery associated with a paleosol that was also covered with pumice, indicating a volcanic eruption. Unfortunately, this pottery was never published and the site is now destroyed.[59]

On the island of Tilos, southeast of Nisyros, tephra layers were found in a trench that also produced pottery from the Middle Minoan, the Late Minoan IA, the Late Minoan IB, and into Late Minoan II.[60] Geologists studying the Tilos tephras have not found any on the island dating to the Bronze Age,[61] and no studies have been done to determine if the tephra deposits associated with the Minoan pottery on Tilos are primary deposits, or reworked/redeposited older deposits.

Although a lack of precision in the dating and the absence of detailed study of the Aegean sites precludes certainty, the various factors mentioned above do suggest that there was an eruption of a volcanic vent in the area of Yali and Nisyros in or around the mid fifteenth century B.C.E. If the vent was above sea level at the time of the eruption it would have produced airborne tephra, and the vent's collapse and submergence into the sea could have triggered one or more tsunamis that caused Deukalion's flood along the coasts of southeastern Greece and across the northern shore of Crete. Tsunamis directed toward the south or southeast diffracting around nearby islands would cross the Mediterranean to the coast of Israel and the Egyptian Delta. Had the local winds been coming from the northwest at that time, ash particles would have blown southeast to Egypt. Since the timing of these events is also unknown, an eruption cloud could have arrived at the Delta before the collapse of the vent and any accompanying tsunamis.

THE MIRACLE AT THE SEA

The El Arish inscription and the Exodus 14 account both describe an intense darkness. In the biblical account the darkness has been transformed into the pillar of cloud that led the first group of Israelites to the Mountain of God 178 years before. A clue that this was a quite different cloud is found in the last part of Exodus 14:20, an enigmatic passage that reports that the pillar of cloud and fire did not turn to fire on that particular night but remained dark.[62] The El Arish text describes the sort of darkness produced by a tephra cloud. Not as long-lasting or as extensive as the Santorini tephra cloud, this one was nonetheless unexpected and extraordinary.

To the Israelites, camped by Pi-hahiroth awaiting the Egyptian onslaught, the tephra cloud would have been a manifestation of YHWH's divine presence, the great and wondrous cloud in which God resided, coming once again to help them. To the Egyptians, it could only have embodied terror and horror.

Ancient Egyptians believed that each night the sun god Re was opposed by the forces of chaos, embodied in the serpent Apophis who attempted to stop Re from reaching the eastern horizon and starting another day. On sunless days, Egyptian believed, Apophis had, at least temporarily, vanquished Re. For them, unusual darkness, such as a solar eclipse, was terrifying.[63] The longest-reigning Hyksos ruler had the same name as the Egyptian serpent of chaos and, as suggested by the El Arish text, Asiatics were the children of Apophis—in the popular Egyptian mind both the Hyksos ruler and the serpent of chaos. What then, did the Egyptians think, when they awoke at what should have been dawn the next morning to attack the "children of Apophis," this band of recalcitrant Asiatic slaves, and saw nothing but the darkness of the tephra cloud? To them, the serpent of chaos, Apophis, had defeated the sun god Re! The panic among the Egyptian horses and men must have been enormous (see Exodus 14:24b), and they probably balked at moving out. Exodus 14:25b says: "Let us flee from the Israelites, for the Lord is fighting for them against Egypt."

Despite the panic among his troops Pharaoh was able to lead his men and chariots out of the fort across the bridge or dike that linked Hebua I to the small peninsula on the other side of the Shi-Hor (see figure 10.1). A strong military leader, such as Tuthmosis III must have been, could have regained control of at least some of his troops and with these advanced on the Israelites.

The tsunami waves, traveling south from the Aegean, would have gained height as they reached the shallow water off the Delta. Crashing through the breaks in the kurkur ridge they would have expanded across the coastal Shi-Hor lagoon in all directions. Roaring across the shallow lagoon, they engulfed the Egyptians crossing the narrow peninsula opposite Tjaru, drowning them (Exodus 14:28, 15:4–5, 10).

The Israelites, camped inland and out of the direct line of the waves, were spared. Later, when the drowning of the Egyptians at the land crossing between the two halves of the Shi-Hor lagoon became fused with the older Exodus story of the dry corridor between the two Bitter Lakes or the northernmost extension of the Red Sea, the single wall of water (the tsunami) became two walls of water, one on each side of the dry corridor (see Exodus 14:16). After the wave(s) departed, the Israelites may have collected the Egyptians' weapons, as Josephus reported, and fought any remaining Egyptians.[64] The El-Arish text mentions a fight.

After it was all over, the Egyptians would have concentrated on finding the body of their Pharaoh, Tuthmosis III, for his death meant a victory of the forces of chaos over the forces of order (*maat*) in the universe. Only after the king was buried with proper rites was the cosmos set in order again.[65] That would explain why the queen was at Pi-Kharoti—like the Egyptian goddess Isis, she was looking for the body of her dead husband.[66]

But she may not have found it. The biblical account (Exodus 14:27–28) implies that Pharaoh was drowned with his army. Tsunamis typically suck many of the bodies of their victims back into the sea with the retreating wave, as the Egyptian name of this location, Place of the Whirlpool, implies happened here, although some bodies must have been left on the shore (Exodus 14:30*b*). In 1881 Egyptologists discovered a cache of royal mummies at Deir el-Bahri near Thebes, and in 1898 the tomb of Amenophis II in the Valley of the Kings yielded a second group of mummified royal remains. The mummies in both caches, individuals from the Seventeenth through the Nineteenth Dynasties, had been stripped and mutilated by ancient tomb robbers. During the Twenty-First Dynasty, after being moved several times, the mummies were rewrapped and relabeled.[67] Among those identified by their Twenty-First Dynasty labels were Tuthmosis I, Tuthmosis II, Tuthmosis III, Amenophis II, Tuthmosis IV, and Amenophis III.

Starting in 1967 these royal mummies, now at the Cairo Museum, were X-rayed by a University of Michigan–University of Alexandria team headed by James E. Harris, then Chairman of the Department of Orthodontics at the University of Michigan. Edward F. Wente of The University of Chicago's Oriental Institute was also called in to provide the historical ages and family trees. Based on X-rays of the skeletons, physical anthropologists provided detailed estimates of the ages of the mummies at death.[68] Later, Harris went on to analyze the craniofacial bones of each mummy with detailed computer imaging and statistical analyses of 177 data points from each skull. Given that these craniofacial features are the result of inheritance, his analyses provide the best available approximations of genetic relationships and can either affirm or contradict the labeled identities of the mummies.[69]

The age estimates based on analyses of the X-rays were generally and often markedly younger than the agreed-upon historical ages of many of the mummies. For example, the body in the coffin of the presumably middle-aged Tuthmosis I was that of a youth of eighteen to twenty-two years who did not have the pharaonic pose of arms crossed over the chest. Harris' craniofacial analyses highlighted even more mismatches. The mummy labeled as Seti II of the Nineteenth Dynasty "bears a striking resemblance to the kings of the Eighteenth Dynasty" (he is probably Tuthmosis II, and the mummy labeled Tuthmosis II is Tuthmosis I). The craniofacial morphology of the supposed Amenophis II mummy "does not suit his being the son of Thutmose III and father of Thutmose IV." In fact, the Amenophis II mummy is the only one suitable to be the father of the Amenophis III mummy, who strikingly resembles images of Akhenaten (Amenophis IV).[70]

The mummy supposed to be Tuthmosis III was identified by a shroud folded on top of the mummy, which was found in the now-stripped outer coffin of that pharaoh. It is, however, estimated to be thirty-five to forty years of age at death, and Tuthmosis III was about sixty years of age when he died. According to Edward Wente, James Harris' analyses suggest that the most likely genealogical sequence *for the mummies* is: Tuthmosis III, Tuthmosis IV, Amenophis II, and Amenophis III.[71] Tuthmosis IV is the only one of this group that was correctly identified by the Twenty-First Dynasty restorers. Thus the mummy labeled Tuthmosis III is more likely to be that of Amenophis II, Tuthmosis III's son. The age is a better fit as well, since this mummy's estimated age (thirty-five to forty years) is close to the estimated historical age of Amenophis II at death (forty-four years).[72] If Harris and Wente's reidentifications are correct, the real possibility exists that the mummy of Tuthmosis III has not been found.[73]

RETURN TO CANAAN

While the Egyptians were busy collecting their dead, the fleeing Israelites would have turned south, away from the Way of Horus (called "the way of the land of the Philistines" in the Bible—an anachronism). Instead, they made their way across the Sinai along the "Way of Shur" (see figure 5.2 and Exodus 15:22a: "and they went into the Wilderness of Shur") that led eventually to Beer-sheba and Hebron.[74] Reunited with their own clan and tribal groups at last, they told the story of their struggles with Pharaoh and their miraculous deliverance: their covenant slaughter-offering to YHWH, the disease that plagued the Egyptians but passed them over, their expulsion by Pharaoh, and finally their rescue by YHWH and the death of the Egyptians by the waters of the Reed Sea. This story became a living and vital part of their tribal traditions, joining the tradition stories of the earlier exodus in the composite lore of these peoples. This would have a considerable effect on the tradition history of the Israelites. As the years and then the centuries passed these two sets of tradition stories would fuse or merge, as similar stories invariably do in oral traditions, into one composite epic story of one Exodus from Egypt, as we will see in the next chapter.

CHAPTER ELEVEN

The Formation of the Exodus Tradition

In the two and a half centuries following the second exodus, Israelite tribal groups continued to live as pastoralists in the highlands and along the edges of the desert areas of Canaan. For them, virtually all of this period fell into what oral historian Jan Vansina termed a "floating gap" (see the Introduction and Appendix) and, once forgotten, could not be reclaimed in the historical record. Only the more notable tenures of certain judges such as Gideon and Deborah were remembered from this period. Biblical scholar Frank Moore Cross has suggested that during this period of Israelite history many of the elements of tribal folklore were incorporated into a larger epic tradition that was recited each year at the spring covenant festivals of the tribes. He notes that the two cores of this epic tradition are the divine victory at the sea and the covenant at Sinai.[1] In fact, these two themes reflect the two exodus stories, and it was during these centuries that the two exoduses merged into one grand, all-encompassing Exodus tradition.

In the remainder of the Late Bronze Age after the second exodus (from 1450 to about 1200 B.C.E.), the earlier exodus under Moses at the time of the Minoan eruption of Santorini and the ensuing journey to the Mountain of God became merged, at least for those groups whose members had participated in the subsequent event, with the story of the mid-fifteenth century exodus at the time of the second volcanic eruption offshore of Yali and the accompanying destruction of the Egyptian army by the shores of the sea. As in most cases of memory fusion, the more recent event (the fifteenth century B.C.E. exodus) overshadowed the older set of memories, except where the older story contained elements not found in the more recent one. The first nine plagues and Moses, the overwhelming leader, as well as the journey to the Mountain of God and the destruction of Jericho had no counterparts in the later exodus story and so survived relatively intact. Elsewhere, the framework of the more recent exodus dominated. Some of the most important elements that merged are shown in table 11.1.

Rather than being farmers and pastoralists in the Wadi Tumilat under a Hyksos overlord, as they were at the time of the first exodus, in the final tradition story the Israelites were slaves in one or more store cities of the Egyptian Delta, making bricks for a hard-hearted, native Egyptian pharaoh, as they were at the time of the second exodus. The anonymous negotiators at the time of the second exodus were replaced by Moses, the leader of the first exodus (with the later addition of Aaron). Pharaoh Tuthmosis III retained his overbearing personality but lost his name in later Israelite tradition, a common occurrence since names often drop out in the leveling process (see Appendix). Only the Egyptians, as recorded by Manetho (see chapter 10), seem to have retained the monarch's name in their own traditions of these events.

The time of the first exodus, at the beginning of February, was lost as the story was shifted to the spring equinox, the approximate time of the second exodus. This shift made sense to later Israelites in Canaan because the original story included the harvest times for the barley and wheat. In Canaan the barley was harvested beginning in late March and the wheat in late May–early June. These later Israelites needed to make sense of these harvest times, and they did not realize that the harvest times in the Egyptian Delta were about two months earlier than harvests in Canaan.

This nearly two-month shift caused the first exodus's Feast of Unleavened Bread to become merged, albeit incompletely, with the second exodus's Passover. The Passover, which in historical fact had been the annual commemoration and renewal of their people's first, seventeenth century B.C.E., covenant sacrifice and meal at the Mountain of God, thus became inextricably linked with the second, fifteenth century B.C.E. exodus.

The events at the time of the two volcanic eruptions on the Aegean Volcanic Arc also merged. The nine plagues caused (except for the locusts) by the Santorini eruption's tsunamis and ashfalls merged with the death of the noninfant children at the time of the second exodus, for a total of ten plagues in the composite exodus story. The fifteenth century ashfall and tsunamis from the volcanic eruption off Yali, saving the Israelite slaves from Pharaoh and his army at the shores of the Mediterranean Sea (just north of the Ballah Lakes, the Reed Sea), was merged with the seventeenth century B.C.E. story of the passing of the Israelites from the Wadi Tumilat

TABLE 11.1.
Comparison of the Exoduses

	First Exodus	Second Exodus	Composite Exodus in Israelite Tradition
Date	1628 B.C.E.	About 1450 B.C.E.	480 years before the building of Solomon's Temple (1 Kings 6:1)
Time of year	Beginning of February	Shortly before the spring equinox	Shortly after the spring equinox
Commemorative event	Baking and eating of unleavened bread	Annual covenant sacrifice and meal	Festival of Unleavened Bread and the Passover
Israelite leader	Moses	Elders among the Israelite captives in Egypt	Moses (Aaron is added later)
Egyptian leader	Unknown	Pharaoh Tuthmosis III	Pharaoh
Plagues	Nos. 1–9, with all but the locusts (no. 8) caused by Santorini tsunami and ashfall on the Delta	No. 10, death of noninfant children probably caused by food poisoning	Nos. 1–10 (all plagues become merged)
Type	Flight	Expulsion by Tuthmosis III	Expulsion by Pharaoh
Pursuit	No	Yes	Yes
Pillar of cloud and fire	An erupting Arabian volcano visible to the Israelites only as they neared it in Arabia, *not in Egypt*	Ash cloud from an erupting volcano near the island of Yali blown Southeast to the Nile Delta; seen there but *not followed by anyone*	Pillar of cloud and fire followed by the Israelites out of Egypt to the Mountain of God (Mount Sinai)
Events at crossing	Land ridge exposed by wind, allowing dry passage for Israelites and their animals	Darkness from ashfall; some Egyptians in chariots drowned by tsunami; possibly a fight	Pillar of cloud; waters part, allowing Israelites dry passage; walls of water return, drowning Egyptians in chariots

(Continued)

TABLE 11.1. (*Continued*)

	First Exodus	Second Exodus	Composite Exodus in Israelite Tradition
Crossing place into the Sinai Peninsula	Land ridge between the Bitter Lakes or the northernmost extension of the Red Sea	Close to the Mediterranean coast near the Ballah Lakes (the Reed Sea)	At the Red or Reed Sea
Route across the Sinai and destination	Probably the Way of Seir to the Gulf of Aqaba and to Arabia	Way of Shur back to Canaan	Unclear, but Wilderness of Shur is mentioned, to the Mountain of God

into the Sinai Peninsula across a land ridge between the Bitter Lakes or the Red Sea. However, the traditional time interval between the Exodus and the giving of the covenant at Sinai was retained in Israelite tradition. This resulted in the emergence of an association of the giving of the Torah at the Mountain of God with the Festival of Shavuot (Festival of Weeks), which marks the end of the barley harvest and the beginning of the wheat harvest *in Canaan*.

EGYPT IN CANAAN 1450–1200 B.C.E.

There is no mention of an Egyptian presence in Canaan in the biblical narratives of the tribal or judges' period of Israel's history. The only hints in the biblical texts of Egyptian contacts come in the names Rameses and Pithom for the Delta localities occupied at earlier times by the Israelites: Avaris and *T̠(k)w*. During this period nomadic shasu watered their flocks near Pithom and, in all likelihood, later Israelite slaves labored at Pi-Ramesses, capital city of the Nineteenth Dynasty's greatest ruler, Ramesses II. Escapees or travelers would have carried these names back to Canaan where they were incorporated into Israelite tradition by storytellers explaining to their listeners what these places were called in their own time.

During the earlier part of the Nineteenth Dynasty (thirteenth century B.C.E.), the Sinai Peninsula forts along the Way of Horus were enlarged or rebuilt and manned with Egyptian garrisons, while certain towns in Canaan along the principal trade routes acquired Egyptian administrators and garrisons.[2] Nineteenth Dynasty control of the isthmus of Suez and of the land route across the northern Sinai would have precluded the escape of all but a few individual slaves to Canaan, however. One late Nineteenth Dynasty Egyptian papyrus contains an account of only *two* slaves escaping through the lakes of the isthmus and their pursuit by Egyptian authorities.[3] It is clear from this text that even so few escapees called for extensive actions by the Egyptians guarding the border. A large group of slaves, such as the Israelites, fleeing successfully across the Sinai is not historically feasible during this time period.

ISRAELITE POPULATION GROWTH AND SETTLEMENT

By the end of the Late Bronze Age (ca. 1200 B.C.E.) the Israelites living in the central highlands (the hill country Ephraim and Manasseh) of Canaan had become "a numerous people" (Joshua 17:17). The need to feed their growing population now led them to begin settling down in permanent villages to supplement meat and milk from their herds with the cultivation of cereal grains. The first Israelite villages, in the late thirteenth century B.C.E., were near highland Canaanite cities where the land was already deforested, and also on the desert fringe.[4] Later, settlement expanded westward into the forested areas of the highlands (see Joshua 17:15) and out onto the western slopes where the available agricultural land was maximized through the building of terraces.[5] From Ephraim and Manasseh Israelite settlement spread north into Lower Galilee in the twelfth century. In Upper Galilee, the tribe of Naphtali crystallized in the territory of Hazor and the tribe of Asher within the territory of Acco, probably after these cities had been destroyed in the thirteenth century B.C.E.[6]

In the hill country of Judah, the archaeological evidence suggests that there was a transition from nomadic to settled life beginning in the second half of the thirteenth century, and that most of the Iron Age I settlers were originally members of local pastoral groups. An exception is in the area of Jerusalem, where northern peoples (the Jebusites) conquered the city and its surrounding territory.[7]

The central hill-country population in the thirteenth century B.C.E. (toward the end of the Late Bronze Age) is estimated at about twelve thousand people. By the twelfth century, it had grown to about fifty-five thousand people and by the eleventh century B.C.E. to about seventy thousand.[8] Renee Pennington's analyses of early human population growth rates demonstrate that these increases are in line with the initial Israelite nomadic population (seven thousand to ten thousand) in the Late Bronze

Age, particularly because there are "substantial increases in the survival of young children as populations switch from nomadic to sedentary lives," even when, as typically happens when people settle down, the overall death rate increases.[9] As mentioned in chapter 9, survival of young children is far and away the most important factor in overall population growth.

CLIMATE CHANGE AT THE END OF THE LATE BRONZE AGE

At the same time that the Israelites started becoming sedentary villagers— late in the thirteenth century B.C.E.—the climate in Europe and Africa changed from cool and wet to warmer and dryer. This change, in the last part of what is called the Subboreal phase in Europe, was recorded in Alpine lake levels and in atmospheric ^{14}C variations and was connected to fluctuations in the Sun's radiation.[10] The Sahara began to get dryer again after 3000 B.P. (about 1250 B.C.E.); in East Africa, lake levels fell dramatically beginning in 1260 ± 50 B.C.E. and agriculture ceased in Nubia after the reign of Ramesses II.[11] In Canaan, the Dead Sea fell to twenty to twenty-five meters lower than it had been during the Late Bronze Age.[12] Soils in coastal Syria (Ugarit) and Cyprus reflect hotter and dryer climatic conditions at this time.[13] Analysis of modern-day rainfall and drought patterns shows that a pattern that produced drought through most of Greece and the Near East fits well with that of the supposed population movements at the end of the Late Bronze Age.[14]

With this climatic change, the delicate balance between nomadic pastoralists and farmers in the Near East was disturbed. Beginning at this time, texts report nomadic incursions and migrations, disastrous famines and droughts in Anatolia, Syria, Mesopotamia, and Libya lasting into the tenth century B.C.E., and evidence of low Nile flood levels starting in the second half of the thirteenth century B.C.E.[15] The pharaoh Merneptah (1212–1202 B.C.E.) sent grain to the Hittites in Anatolia in 1212 to relieve a famine, but decreased Egyptian agricultural output caused by the generally worsening climatic conditions must have been a decisive factor in instituting a new Egyptian policy toward Canaan late in the thirteenth century B.C.E.

FROM BRONZE AGE TO IRON AGE IN CANAAN

Described on Cairo Museum Stela No. 34025 and depicted on a series of reliefs in the temple of Karnak in Thebes is a campaign in which Merneptah captures the Canaanite cities of Ashkelon, Gezer, and Yano'am and lays waste to "Israel"—not a town but a people, with an eponymous male ancestor. Merneptah (or his son) also captured some shasu nomads and brought them back to Egypt with the captive Canaanites. The men of "Israel" depicted in the Karnak reliefs are dressed like other Canaanites in long gowns; the shasu wear kilts. The town-dwelling Canaanites are depicted as having chariots, possibly the "chariots of iron" referred to in Judges 1:19. The "Wells of Merneptah" referred to in Papyrus Anastasi III may have been the same place name as the well of waters of Nephtoah in Joshua 15:9 and Joshua 18:15, the proper name of the pharaoh having passed into the Hebrew and become garbled in its oral transmission through the centuries.[16]

Whether Merneptah's campaign was in response to a Canaanite rebellion against the tighter controls the Egyptians were putting into place, or itself the initiation of these controls, its net effects were direct Egyptian rule in a number of areas of Canaan, more Egyptian taxation, and the building of Egyptian administrative centers throughout the country.[17] One of its most significant consequences was that large quantities of Canaanite grain passed to direct Egyptian control.[18]

All pastoral groups, including the Israelites of the Canaanite highlands and desert fringe, are dependent upon their settled neighbors for grain, pottery, and other manufactured products. Because of this dependency, the Egyptian sequestration of Canaanite grain stores beginning in the late thirteenth century B.C.E. and the destruction of many lowland cities in the thirteenth and twelfth centuries would have exerted enormous pressure on the highland Israelites to accelerate their already-begun process of settling down to village life.[19] The archaeological evidence from Iron Age I villages does show certain features that indicate recent sedentization: a site layout that is a series of connected rooms forming an oval around a central open courtyard, reminiscent of a nomadic tent encampment; the so-called four-room pillared house (with stabling for animals along the sides of a main room); and large stone-lined silos (usually one to two meters in diameter), typical of the way many newly sedentary people store their grain.[20]

Highland pottery from this early Iron Age I period, although mostly an assortment of large, rough, undecorated vessels, also contains some styles that clearly harken back to the earlier Late Bronze Age inhabitants of Canaan, suggesting that the majority of early Iron I villagers had been in the area for some time. This mixture of old and new is most evident in the hill country of Manasseh, but existed as far north as Galilee.[21] This combination of old and new accords well with the notion that the Israelites had been nomadic occupants of Canaan during the Late Bronze Age, and then formed the majority of the settlers in the Iron I highland villages and hamlets. As "local" pastoralists they would have used local Canaanite pottery styles; as newly sedentary people (and ethnically distinct from the urban Canaanites) they would have their own house types, settlement layouts, and storage pits.[22]

EARTHQUAKES AND THE MIGRATIONS OF THE SEA PEOPLES

In addition to the climate changes in the late thirteenth century B.C.E. there is physical evidence of a sequence of earthquakes in the Aegean and Anatolian area between 1225 and 1175 B.C.E., a series of disasters that devastated many Late Bronze Age cities and towns. These earthquakes, occurring along one or more connected tectonic faults as strain passed from one section of a fault to the next, with one earthquake on the fault triggering another one in an adjacent section, have been termed an "earthquake storm."[23] Together, these climatic and tectonic events spelled the end for the Hittite Empire in Anatolia and set in motion a series of migrations of peoples from the Aegean and Anatolian areas of the eastern Mediterranean.

For the peoples of Canaan, the most important of these migrations was that of the "Sea Peoples" who invaded Egypt during the reign of Pharaoh Ramesses III (1184–1153 or 1151 B.C.E.). Some of these invaders appear in the Bible as the Philistines. Recent scholarship has been able to show that they were related to the Mycenaeans of Greece.[24] According to Egyptian sources, the pharaoh repulsed the invaders in 1175 and settled some of them on the coast of southern Canaan. More recent work suggests that they had already invaded southern Canaan prior to the battle with the Egyptians, exterminating the populations of several Canaanite cities (see figure 9.1). Rather than actually controlling the Philistines, the Egyptians were merely able to contain them to a narrow coastal strip about twenty kilometers wide and fifty kilometers long until about 1150 B.C.E.[25]

Battered by the Sea Peoples' invasion and continuing poor harvests—after 1170 B.C.E. grain prices in Egypt increased to eight times their earlier amount, peaking at 1130 and stabilizing about 1110 B.C.E.[26]—the Egyptians withdrew from their cities in Canaan in the late twelfth century B.C.E. Following this withdrawal the Philistines expanded in all directions.[27] Soon they came head to head with the Israelites expanding westward across the highlands and into the foothills.[28] In the battle of Ebenezer, described in 1 Samuel 4:1–11, the Philistines defeated the Israelites, captured the ark, and destroyed the cult center at Shiloh. The break in the Israelite year count, which Josephus incorrectly identified with the commencement of the first Temple in Jerusalem, probably marks this disaster instead.[29] As such the defeat and destruction of Shiloh would have occurred in 1056 B.C.E.

FROM TRADITION TO PROTO-HISTORY

The generalized weakness of the Egyptian, Assyrian, and Babylonian empires during this period of drought and poor harvests, the shift of the Israelites from small nomadic groups to larger and more politically sophisticated sedentary units, coupled with the need for the various Israelite tribes to unite under a strong leader to defend themselves against the invading Philistines, provided the impetus for the creation and the sustaining of the Israelite monarchy.

The beginning of the monarchy is the time, according to Abraham Malamat, in which Israel entered its "historical" period.[30] More realistically, this period is better termed "proto-historical," a time during which the traditional stories of the tribal league became the traditions of the United Monarchy, traditions which eventually found their way into the early written texts that were the precursors of the later biblical texts we have today.

RECREATING THE EVENTS BEHIND THE EXODUS TRADITION

This book has shown how natural phenomena are connected with the biblical accounts of the Exodus, the Sojourn in the Wilderness, and the Israelites' conquest of Canaan. It differs from previous books and articles about the connection between the Exodus and a volcanic eruption in that there are *two* volcanic eruptions and *two* exoduses related to the Exodus found in the Bible: the first, the 1628 B.C.E. Minoan eruption of Santorini/Thera that was the cause of most of the Exodus plagues and the impetus for the first departure from Egypt and the second, an eruption from a volcanic vent off the Aegean island of Yali 178 years later that caused another period of darkness and a series of tsunamis that drowned the pursuing Egyptians during a second exodus from the Egyptian Delta. No one has proposed a composite, two-part Exodus with two separate volcanic eruptions, and no one has pointed out the connection between the eruption off Yali and a second exodus. This hypothesis also explains many of the inconsistencies in the Exodus story that previous hypotheses have not been able to (or ignored entirely): the request by Moses (or the Israelite elders) to go on a three days' journey into the wilderness to give a slaughter-offering to their god, the seeming separation of the Feast of the Unleavened Bread and the Passover, the burning bush and the obviously volcanic description of the Mountain of God, and the timing of the biblical events in relation to the destruction of Jericho and the conquest of Canaan.

No one, to my knowledge, has taken seriously the dates given by the first century Jewish historian Josephus and found in them not only a near-accurate date for the first exodus but also the correct time of year for it. With this knowledge, and with an accurate scientific date for the destruction of Jericho, which is so vividly and correctly described in the biblical text, we can go back to the correct time period for the first exodus, during the Hyksos rule of Egypt. Archaeological evidence from the Wadi Tumilat for that period clearly shows a distinct sub-group of Semitic pastoralists who suddenly and inexplicably deserted the Wadi and never returned, at about the time of the Minoan eruption of the Santorini/Thera volcano.

The latest scientific information on the Minoan eruption conforms *in detail* to the description and order of eight of the first nine Exodus plagues. The biblical story of the unleavened bread is a reliable time marker showing that this exodus occurred before Egypt's New Kingdom (which began about 1550 B.C.E.). Other time markers show that a second exodus occurred during the New Kingdom, when native Egyptian rule was restored.

By looking at an early Egyptian tradition about Moses, one discounted by nearly all scholars, we can see that part of the Moses story related to the Thirteenth Dynasty ruler Khaneferre Sobekhotep IV and his contemporary, the Fourteenth Dynasty ruler Nehesy. Names in the family genealogy of Moses relate to this story and to Nehesy himself. This part of the Moses story actually relates to an earlier Moses, for whom the biblical Moses was named.

The biblical Moses, the Moses of the first exodus, fled to Midian where he saw what was most likely a volcanic fire fountain, described in the biblical tradition as the burning bush. Returning to the Wadi Tumilat he was able, because of the plagues precipitated by the Minoan eruption, to persuade his people to accompany him to the Mountain of God and make a new covenant there with the god of their ancestor Abraham. One of the most important new findings presented in this book is that Moses and his people arrived at the Mountain of God at the very beginning of spring, and they held their covenant offering and feast at the *first full moon* after the spring equinox. This, of course, is when the Passover is held.

After a failed attempt to invade Canaan the Israelites spent seventy-eight years, not forty, in the wilderness, or into their fourth generation. Moses died during this time, but earlier he had formed an alliance with the Kenites under their leader Hur. The rebellion under Aaron, a fellow Levite, destroyed the first set of covenant tablets, but it failed because of Levite support for Moses, and Aaron was executed at the Mountain of God. Eventually global and regional climatic changes sparked the emergence of bubonic plague in Egypt, helping the Egyptians to defeat the Hyksos. The flight of the Hyksos back to Canaan spread the plague to Jericho and also to the Midianites and Israelites camped east of the Jordan River.

At this time an earthquake dammed the Jordan River, allowing the Israelites to pass dryshod across to the western side of the river not far from Jericho. An aftershock destroyed the already weakened walls of Jericho while the Israelites were besieging it, and the town was destroyed. The purification rites described in the Bible after the Israelite conquest of Jericho closely resemble measures taken to avoid contamination by the bubonic plague.

The bubonic plague was a hitherto unsuspected factor in the fall of Canaanite cities and towns at the end of the Middle Bronze Age, and malaria played a role in the population increase of the Israelites relative to their lowland Canaanite neighbors in the Late Bronze Age. An important finding highlighted in this book is that the oldest Israelite tribal boundaries relate to Dead Sea levels that existed before the fifteenth century B.C.E. This eliminates an Israelite conquest of Canaan during or after the time of Ramesses II, thought by many to have been the pharaoh of the Exodus. Various groups, both Israelite and non-Israelite, including one offshoot of the Israelite tribe of Reuben, united to form the biblical tribe of Judah.

Israelite shasu were slaves in Egypt during the reign of Pharaoh Tuthmosis III, the pharaoh of the second exodus. These slaves held the first Passover in about 1450 B.C.E. It was in fact their annual commemoration of the first covenant sacrifice and meal at the Mountain of God in 1628 B.C.E. After the death of numerous Egyptian children, the slaves were expelled but then pursued by Pharaoh Tuthmosis III and a force of Egyptian chariots. This second exodus coincided with another volcanic eruption off the island of Yali in the Aegean Sea. A tephra cloud from this eruption caused the darkness mentioned both in the Bible and in an ancient Egyptian inscription from El Arish, and the tsunamis produced by this eruption drowned Tuthmosis III near the Mediterranean Sea north of the Ballah Lakes (the Reed Sea). His body was probably not recovered and the mummy said to be his is more likely that of his son.

After they had returned to their own tribal groups, the fleeing Israelites told their stories of the remarkable events that had happened. These stories became merged, as the centuries passed, with the stories of the first exodus. By the time of the United Monarchy there was only one Exodus, one Passover, one journey, and one overwhelming leader, Moses. This Exodus tradition became the foundation story of Israel. It still is.

APPENDIX

Oral Transmission, Memory and Recall, and Oral History

ORAL TRANSMISSION

Oral transmission involves the telling of information, starting with the first recounting of the information and continuing through any number of retellings. This transmission occurs in a series of stages, with each successive stage getting farther and farther from the original version. In the first stage are eyewitness accounts, hearsay, and rumor.[1] With time, eyewitness accounts, that is, recollections of past events, become intermingled with hearsay and rumor, as groups of people tell stories to other people. Hearsay and rumor reflect what a group is thinking and feeling; even more than eyewitness accounts, they form the basis of later oral tradition.

STUDIES OF MEMORY AND RECALL

In the early twentieth century a Cambridge University psychologist named Frederic Bartlett undertook a series of experiments about recall. The first set involved having college students read a Native American Indian folktale through twice, then write the story down. Later they were asked to reproduce the folktale after hours, weeks, months, or even years. From these experiments Bartlett concluded that accuracy of remembrance is the rare exception, although "with frequent reproduction the form and items of remembered detail very quickly become stereotyped and thereafter suffered little change."[2] Remembered material is simplified in form, some details forgotten, and others transformed into more familiar forms.

Bartlett's second set of experiments involved transmission of the story from one subject to another through a chain of ten individuals, much like the old party game Telephone. He frequently found that opposition or trans-position took place: incidents and events were transposed, terms changed to their opposites, opinions and conclusions reversed. Names and numbers rarely survived intact.[3]

Bartlett used stories completely unfamiliar to his subjects and did not recreate real-world conditions. However, people usually remember better things that are more familiar to them. Then they don't have to remember exotic names, odd eating practices or rituals, strange scenery, or peculiar social relationships. When such exotica do pop up in a story, they may be holdovers from an original account.

In contrast to individual memories, memories carried down by a group are richer and more detailed, as members of a group remind each other of things. Errors in the sequencing of events are more likely to be corrected by the group, and "false memories" to be cut out of the collective narrative as the group reaches consensus. As one study concluded: "Social recall [that is, group memory] is an improvement on individual performance. It is more accurate, more complete, and produces no decrement in subjec-tive or objective validity."[4] However, "implicational" errors—additions, plausible deductions, interpretations, and explanations, *which do not con-tradict the original version*—are more common in group than in individual recall. In fact, the introduction of implicational errors seems to be an im-portant part of the construction of a group's narrative.[5] Groups try even harder than individuals to provide a coherent narrative, a story that "makes sense" in terms of the tellers and the listeners. This means that when we find concrete items in an oral tradition (or in a narrative that comes from an oral tradition), items that do not agree with the narrative as a whole or are clearly out of place in the structure of the story, they are likely to be holdovers from an original account.

Some events are remembered better than others. Low-frequency events are remembered better than more common occurrences.[6] Also standing out are "landmark events," public and personal landmarks that reduce dating errors. The title of one research study on landmark events de-scribes their significance as well as anything can: "Since the Eruption of Mt. St. Helens, Has Anyone Beaten You Up? Improving the Accuracy of Retrospective Reports with Landmark Events."[7] Without such important landmark events, the passage of time fares poorly in personal and even in group memory.

Generally, the farther back in time an event is, the more poorly it will be remembered, but there are important exceptions to this rule. The most important category of exceptions involves telescoping. Telescoping refers to the systematic errors that result when people's estimates of dates are moved from the actual time of the event *forward toward the present* so that time between the event, and now, becomes compressed, much like a scene viewed through a telescope. Without secure landmarks, human memory will inevitably telescope in some form or another.[8] For example, as you remember the last time you met a good friend who lives out of town, you can recall vividly certain aspects of the meeting and what the two of you said to each other. You think the meeting was about two years ago, but then you find an old letter from the friend mentioning the meeting—and the letter is ten years old! Such well-remembered events are dated in one's memory as more recent than they really are. Since low-frequency events are known to be well remembered, they will be prime candidates for telescoping in a person's or a group's recall.

RUMOR TRANSMISSION

Rumor transmission works in much the same way as does individual and group recall, only faster. Rumor spreads along a well-defined path. The first and most common step in this process is *leveling*, where details are eliminated.[9] Stories grow shorter and more concise as they travel from one person to the next, in order to make the story easier to grasp. Details are lost and the statement as a whole gets more concise. A few minutes of concerted rumor-spreading within a group will accomplish as much information loss as several weeks of forgetting by a single individual. What is left is more likely to be a short, concise statement that remains relatively stable over time. In the leveling process, facts are more likely to be lost in the less important parts of the narrative, and information will simply fall out (be forgotten) rather than be lost to distortion. In fact, distortion of any kind is minimal if an item is important to the people telling it.[10] The middle part of the message will be the least well retained.[11]

After leveling, the next step or stage of rumor transmission involves *sharpening*, or giving selective attention to particular information.[12] Whatever remains of a rumor after leveling will automatically sharpen. Sharpening is exaggeration, enhancement, embellishment. It is one type of implicational error.

The final segment of the path of rumor distortion involves additions or major changes not part of the original material but added to fulfill needs of those transmitting the story. These changes are called assimilation or structuring There are four types[13]: (1) changes (or additions) to fit the main theme of the story; (2) condensation (or fusion), where similar incidents, similar individuals, or other related items are fused into a single incident, individual, or item or, very often, details from the end of a story are fused with those from the beginning; (3) changes to expectation, where items are reported as they are expected to be, not as they originally were; and (4) changes to linguistic habits, where words in the original version are transformed to those more familiar to the subject's speech or knowledge base. For example, in one study conducted in Liverpool, England, the name of the hotel in the test story, the Astoria, was transformed by those recounting the story to the name of Liverpool's principal hotel, the Adelphi.[14] Also, out-of-date words or terms are transformed to more contemporary ones.

VANSINA AND ORAL TRADITION

Working among nonliterate sub-Saharan African societies in the 1950s and 1960s, a Belgian researcher named Jan Vansina found that oral transmission in these groups mirrored the processes discovered by psychologists in their controlled studies. Like Bartlett, Vansina found that incoming events are assimilated into culturally specific "schemata" in a person's memory. Memory, Vansina wrote,[15] is like a library. Events are processed, labeled, and stored. Forgetting occurs when an event is not labeled in the first place, the label is later destroyed, or the event is "misfiled" in the brain's memory. Concrete items and cliches are remembered better than abstract items; thus numbers are remembered poorly, being abstract and having only sequential labels to define them.[16]

Vansina found that, with time, group traditions tend to become shorter and turn into single anecdotes (leveling). Personal memories become combined to reach a common version, selection favors generalization, imagined links between events are inserted (implicational errors), and discrepant recollections are weeded out. Vansina and others also found that a strong leader may sometimes impose his version of events on a group.[17]

Vansina labeled one important phenomenon he found in oral traditions the "floating gap."[18] This is a form of telescoping. Earlier events in a sequence of related events or people are well remembered. Telescoping brings these earlier items forward in time to join recent events or people, which, because they are recent, are also well remembered. As items recede in time, they are replaced by more recent events and thus fall into the gap, which moves or "floats" with the passage of time. This phenomenon is most frequently found in oral genealogies, but can also be pervasive in stories carried down orally through generations, as notable events are telescoped together and more mundane events pass into the oblivion of permanent forgetting.

Abbreviations

ASHL   *The Archaeology of Society in the Holy Land*, edited by
T. E. Levy. London: Leicester University Press, 1995.

BAR   *Biblical Archaeological Review*

BASOR   *Bulletin of the American Schools of Oriental Research*

Exodus   *Exodus: The Egyptian Evidence*, edited by E. S. Frerichs and
L. H. Lesko. Winona Lake, Ind.: Eisenbrauns, 1997.

FNM   *From Nomadism to Monarchy: Archaeological and Historical
Aspects of Early Israel*, edited by I. Finkelstein and
N. Na'aman. Jerusalem and Washington: Israel Exploration
Society and Biblical Archaeology Society, 1994.

GSABull   *Geological Society of America Bulletin*

Hyksos   *The Hyksos: New Historical and Archaeological Perspectives*,
edited by E. D. Oren. Philadelphia: University of Pennsylvania Museum, 1997.

IJH   *Israelite and Judaean History*, edited by J. H. Hayes and
J. M. Miller. London and Philadelphia: SCM and Trinity
Press, 1977.

IEJ   *Israel Exploration Journal*

JBL   *Journal of Biblical Literature*

JGR   *Journal of Geophysical Research*

JEA   *Journal of Egyptian Archaeology*

JNES   *Journal of Near Eastern Studies*

JVGR   *Journal of Volcanology and Geothermal Research*

NATO   *Third Millennium* BC *Climate Change and Old World Collapse*, edited by H. N. Dalfes, G. Kukla, and H. Weiss. NATO
ASI Series I 49. Berlin: Springer, 1997.

PEQ   *Palestine Exploration Quarterly*

Thera III-2   *Thera and the Aegean World III*. Vol. 2, *Earth Sciences*,
edited by D. A. Hardy, J. Keller, V. P. Galanopoulous,
N. C. Flemming, and T. H. Druitt. London: Thera Foundation, 1990.

Notes

~⁀

Introduction

1. Excerpts of William Leeke's account are in David Howarth, *Waterloo: Day of Battle* (New York: Galahad, 1968) and John Keegan, *The Face of Battle* (New York: Viking, 1976), chap. 3.

2. S. Kassin, P. Ellsworth, and V. Smith, "The 'General Acceptance' of Psychological Research on Eyewitness Testimony: A Survey of the Experts," *American Psychologist* 44 (1989): 1089–98. Some new studies challenge this generalization, however: J. Read, D. Lindsay, and T. Nicholls, "The Relation between Confidence and Accuracy in Eyewitness Identification Studies: Is the Conclusion Changing?" in *Eyewitness Memory: Theoretical and Applied Perspectives*, ed. C. Thompson et al. (Mahwah, N.J.: Lawrence Erlbaum, 1998), 107–30; T. Perfect, T. Hollins, and A. Hunt, "Practice and Feedback Effects on the Confidence-Accuracy Relation in Eyewitness Memory," *Memory* 8 (2000): 235–44.

3. J. Vansina, *Oral Tradition as History* (Madison: University of Wisconsin Press, 1985), 10–11.

4. W. Wagenaar, "My Memory: A Study of Autobiographical Memory over Six Years," *Cognitive Psychology* 18 (1986): 225–52; R. Winningham, I. Hyman, Jr., and D. Dinnel, "Flashbulb Memories? The Effects of When the Initial Memory Report Was Obtained," *Memory* 8 (2000): 209–16; H. Ebbinghaus, *Memory: A Contribution to Experimental Psychology* (New York: Columbia University Teachers College, 1913), 76.

5. Numbers tend to acquire qualitative values such as "many," "few," or "perfect:" Vansina, *Tradition*, 132–33. Names usually drop out unless they have a particular interest to the teller: G. Allport and L. Postman, *The Psychology of Rumor* (New York: Henry Holt, 1947), 124–25.

6. J. C. Miller, "Introduction: Listening for the African Past," in *The African Past Speaks: Essays on Oral Tradition and History*, ed. J. C. Miller (Hamden, Conn.: Archon, 1980), 15–16; Vansina, *Tradition*, 177.

7. Allport and Postman, *Rumor*, 103.

8. W. Brewer, "Memory for Randomly Sampled Autobiographical Events," in *Remembering Reconsidered*, ed. U. Neisser and E. Winograd (Cambridge: Cambridge University Press, 1988), 53–54, 58; idem., "The Theoretical and Empirical Status of the Flashbulb Memory Hypothesis," in *Affect and Accuracy in Recall: Studies of "Flashbulb" Memories*, ed. E. Winograd and U. Neisser (Cambridge: Cambridge University Press, 1992), 290–92.

9. A. Lord, *The Singer of Tales* (Cambridge: Harvard University Press, 1960).

10. Vansina, *Tradition*, 13–14, 130.

11. Vansina, *Tradition*, 17,19–20, 160 (quote), 168.

12. S. Niditch, *Oral World and Written Word* (Louisville: Westminster/John Knox, 1996), 120–21; G. Nagy, *Greek Mythology and Poetics* (Ithaca: Cornell University Press, 1990), 41–42.

13. Niditch, *Oral World*, 122–25; Nagy, *Greek Mythology*, 42.

14. F. Cross, Jr. *From Epic to Canon: History and Literature in Ancient Israel* (Baltimore: Johns Hopkins University Press, 1998), 20–51.

15. Vansina, *Tradition*, 17–18.

16. E. Clark, *Indian Legends of the Pacific Northwest* (Berkeley: University of California Press, 1953), 53–55; S. Harris, *Agents of Chaos: Earthquakes, Volcanoes, and Other Natural Disasters* (Missoula: Mountain Press, 1990), 215–16; D. Vitaliano, *Legends of the Earth: Their Geologic Origins* (Bloomington: University of Indiana Press, 1973), 123–25; C. Zdanowicz, G. Zielinski, and M. Germani, "Mount Mazama Eruption: Calendrical Age Verified and Atmospheric Impact Assessed," *Geology* 27 (1999): 621–24.

17. R. Blong, *The Time of Darkness: Local Legends and Volcanic Reality in Papua New Guinea* (Seattle: University of Washington Press, 1982). In many of these traditions, this ash fall was conflated with an earlier ash fall that was nearly 1200 years old—see M. Mennis, "The Existence of Yomba Island near Madang: Fact or Fiction?" *Oral History* 6, no. 6 (1978): 2–81.

18. For an excellent collection and discussion of natural phenomena preserved in oral traditions see L. Piccardi and W. Masse, eds., *Myth and Geology* (London: Geological Society London Special Publication no. 273).

Chapter 1 Dating the Exodus

1. R. Christiansen, "Epic Movie Adventures with Charlton Heston," *Chicago Tribune*, March 11, 2001.

2. F. Yurco, "Merenptah's Canaanite Campaign and Egyptian Origins," *Exodus*, 46–48.

3. Yurco, "Campaign," 46–47.

4. W. Dever, "Is There Any Archaeological Evidence for the Exodus?" *Exodus*, 70, 74–75.

5. D. Redford, "Exodus I 11," *Vetus Testamentum* 13 (1963): 406–13.

6. D. Redford, "Observations on the Sojourn of the Bene-Israel," Exodus, 62.

7. D. Redford, *Egypt, Canaan, and Israel in Ancient Times* (Princeton: Princeton University Press, 1992), 412–13.

8. J. Weinstein, "Exodus and Archaeological Reality," *Exodus*, 93.

9. A. Malamat "The Exodus: Egyptian Analogies," *Exodus*, 22–24.

10. Malamat, "Egyptian Analogies," 16–17.

11. Weinstein, "Archeological Reality," 92–93; Dever, "Evidence for the Exodus," 74.

12. W. Ward, "Summary and Conclusions," *Exodus*, 106.

13. J. Weinstein, "The Egyptian Empire in Palestine: A Reassessment," *BASOR* 241 (1981): 2–5; W. Dever, "Relations between Syria-Palestine and Egypt in the 'Hyksos' Period," in *Palestine in the Bronze and Iron Ages: Papers in Honor of Olga Tufnell*, ed. J. Tubb (London: Institute of Archaeology, 1985), 70; idem., "The Middle Bronze

Age: The Zenith of the Urban Canaanite Era," *Biblical Archaeologist* 50 (1987): 173–75. The lack of late thirteenth century B.C.E. destructions and/or occupation layers in Canaan is summarized in J. Bimson and D. Livingston, "Redating the Exodus," *BAR* 13 (September–October 1987): 40–45.

14. J. Bimson, *Redating the Exodus and Conquest* (Sheffield: Almond, 1981), 79, 102; Bimson and Livingston, "Redating," 45.

15. B. Halpern, "Radical Exodus Redating Fatally Flawed," *BAR* 13 (November–December 1987): 56–61.

16. J. Jack, *The Date of the Exodus: In the Light of External Evidence* (Edinburgh: Clark, 1925), 219; Jack rejected this date, however, in favor of the 480 years in I Kings 6:1.

17. E. Faulstich, "Studies in O.T. and N.T. Chronology," in *Chronos, Kairos, Christos II: Chronological, Nativity, and Religious Studies in Memory of Ray Summers*, ed. E. Vardaman (Macon, Ga.: Mercer University Press, 1998), 99–101, 102–03, 117.

18. Flavius Josephus, *Against Apion* 2:19.

19. Flavius Josephus, *Antiquities of the Jews* 8:61, 10:147, 20:230, 232.

20. The Maricopa Indians of Arizona used wooden calendar sticks with a notch made in the early spring for each year: L. Spier, *Yuman Tribes of the Gila River* (Chicago: University of Chicago Press, 1933), 138–42. These sticks also included marks for significant events, although details about these events or about breaks in the counting were minimal. The Lakota Indians of the Great Plains may have also used a notched stick to calculate their winter year counts: R. DeMallie and D. Parks, "Tribal Traditions and Records," in *Handbook of North American Indians*. Vol. 13, ed. R. DeMallie (Washington: Smithsonian, 2001), 1070.

21. For the destruction of the Temple: A. Malamat, "Last Kings of Judah and Fall of Jerusalem," IEJ 18 (1968): 154–55.

22. For the destruction of Shiloh: I. Finkelstein, *The Archaeology of the Israelite Settlement* (Jerusalem: Israel Exploration Society, 1988), 220–28.

23. D. Henige, *Oral Historiography* (London: Longman, 1982), 103.

24. G. Hort, "The Plagues of Egypt," *Zeitschrift für die alttestamentliche Wissenschaft* 69 (1957), 92–94.

25. Hort, "Plagues," 95–103 and idem., *Zeitschrift für die alttestamentliche Wissenschaft* 70 (1958), 48–54.

26. M. Krom et al., "Nile River Sediment Fluctuations over the Past 7000 Yr and Their Key Role in Sapropel Development," *Geology* 30 (2002): 71–74; A. Foucault and D. Stanley, "Late Quaternary Palaeoclimatologic Oscillations in East Africa Recorded by Heavy Minerals in the Nile Delta," *Nature* 339 (1989): 44–46. Lower rainfall in Ethiopia leads to less vegetation cover, and more erosion, so that more sediments are washed into the river, which because of the lower rainfall will be flowing (and flooding) at lower levels.

27. J. Garstang and J. B. E. Garstang, *The Story of Jericho* (London: Hodder & Stoughton, 1940), 160–163.

28. P. Francis, *Volcanoes: A Planetary Perspective* (Oxford: Clarendon Press, 1993), 366–67; Smithsonian Institution, *Global Volcanism Program: Volcanic Activity Reports*, www.volcano.si.edu/gvp/volcano/region02/africa_e/nyamura/var.htm and . . . africa_e/nyiragon/var.htm.

29. T. Simkin and L. Siebert, *Volcanoes of the World* (Tucson, Az.: Geoscience Press, 1994), 48–49, note that the Saudi Arabian volcanoes are rift-influenced and have no known eruptions of substantial size. The two northernmost Arabian volcanic areas have had historic eruptions with a low V.E.I. (Volcanic Explosivity Index) of about 2. For more about the Index and the Saudi Arabian volcanoes, see chapters 3 and 5 and the Glossary.

30. A. Galanopoulos, " Die ägyptischen Plagen und der Auszug Israels aus geologischer Sicht," *Das Altertum* 10 (1964): 131–37.

31. G. Davies, "The Wilderness Itineraries and Recent Archaeological Research," in *Studies in the Pentateuch*. Vetus Testamentum Supplement 41, ed. J. Emerton (Leiden: Brill, 1990), 163–66.

32. Vitaliano, *Legends of the Earth*; I. Wilson, *Exodus: The True Story Behind the Biblical Account* (San Francisco: Harper, 1985). E. W. Barber and P. T. Barber, *When They Severed Earth From Sky: How the Human Mind Shapes Myth* (Princeton: Princeton University Press, 2004), especially 80, 85.

33. H. Shanks, "The Exodus and the Crossing of the Red Sea, According to Hans Goedicke," *BAR* 7 (September–October 1981): 42–50.

34. H. Goedicke (translator), "Hatshepsut's Temple Inscription at Speo Artemidos," *BAR* 7 (September–October 1981), 49.

35. A. Gardiner, "Davies' Copy of the Great Speos Artemidos Inscription," *JEA* 32 (1946): 43–56.

36. D. Redford, "Textual Sources for the Hyksos Period," *Hyksos*, 17.

37. H. Goedicke, "The Chronology of the Thera/Santorin Explosion," *Ägypten und Levant* 3 (1992): 57–62.

38. F. Griffith, "The Antiquities of Tell el Yahudiyeh," *Egypt Exploration Fund Memoir 7* (London: Kegan Paul, 1890): 70–74 contains a description of the *naos* and an English translation. For a French translation: G. Goyon, "Les Travaux de Chou et les Tribulations de Geb," *Kêmi* 6 (1936): 1–42. Goedicke's interpretation is in: "Thera/Santorin Explosion," 61.

39. H. Bruins, and J. van der Plicht, "The Exodus Enigma," *Nature* 382 (1996): 213–14. Wiggle-matched to the 1993 tree ring curve there is a 67 percent probability that the true date of the cereal grains is either in the 1601–1566 or the 1561–1524 B.C.E. range. The figures quoted in chapter 8 use a slightly different computer program and the improved 1998 tree ring curve, but both sets of figures are broadly mid (or mid to early) sixteenth century B.C.E. These are high-precision dates from the laboratory at Groningen University. An earlier radiocarbon sample from the Jericho destruction level (BM-1790), was inaccurately dated: S. Bowman, J. Ambers, and M. Leese, "Re-evaluation of British Museum Radiocarbon Dates Issued between 1980 and 1984," *Radiocarbon* 32 (1990): 74. The reevaluated BM-1790 date is 3300 ± 110 radiocarbon years Before Present, in close agreement with the Groningen high-precision dates.

Chapter 2 The Coming of the Hyksos

1. M. Bietak, "The Center of Hyksos Rule: Avaris (Tell el-Dab'a)," *Hyksos*, 91.

2. Flavius Josephus, *Against Apion* 1:75–76; W. Waddell, *Manetho* (Cambridge: Harvard University Press, 1940), 79.

3. M. Bietak, *Avaris and Piramesse: Archaeological Exploration in the Eastern Nile Delta*, Proceedings of the British Academy 65 (Oxford: Oxford University Press, 1979): 226–28.

4. P. Newberry, *Beni Hasan*, Vol. 1 (London: Archaeological Survey of Egypt, 1893), plates 28, 30–31; Bietak, "Hyksos Rule," 99–100.

5. Bietak, "Hyksos Rule," 98–99; J. Holladay, Jr., "The Eastern Nile Delta during the Hyksos and Pre-Hyksos Periods: Toward a Systemic/Socioeconomic Understanding," *Hyksos*, 184.

6. The MBI period alluvial sediments in Nahal Lachish (laminated channel sands, sandy silt levee deposits, and a clayey silt floodplain facies) are followed by a period of erosion in MBII through Late Bronze Age times: A. Rosen, "Environmental Changes and Settlement at Tel Lachish, Israel," *BASOR* 263 (1986): 56–57.

7. For the high Niles of the Twelfth and Thirteenth Dynasties: B. Bell, "Climate and the History of Egypt: The Middle Kingdom," *American Journal of Archaeology* 79 (1975): 223–69. For the correlation between the more northerly monsoons and Mediterranean westerlies: A. Issar and M. Zohar, *Climate Change: Environment and Civilization in the Middle East* (Berlin: Springer, 2004), 67, 153. For modern correlations of high Niles with "La Niña" cold phases: A. Awadallah and J. Rousselle, "Improving Forecasts of Nile Flood Using SST Inputs in TFN Model," *Journal of Hydrologic Engineering* 5 (2000): 377. For the corresponding decrease in precipitation in Israel: D. Yakir, S. Lev-Yadun, and A. Zangvil, "El Nino and Tree Growth near Jerusalem over the Last 20 Years," *Global Change Biology* 2 (1996): 97–101; C. Price et al., "A Possible Link between El Niño and Precipitation in Israel," *Geophysical Research Letters* 25 (1998): 3964.

8. Bietak, "Hyksos Rule," 100–104.

9. Bietak, "Hyksos Rule," 105.

10. Holladay, "Nile Delta," 185; Bietak, *Avaris and Piramesse*, 234.

11. Bietak, "Hyksos Rule," 105.

12. D. Redford, "Textual Sources for the Hyksos Period," *Hyksos*, 4; K. S. B. Ryholt, *The Political Situation in Egypt during the Second Intermediate Period, c. 1800–1550 B.C.* Carsten Niebuhr Inst. Pub. no. 20. (Copenhagen: Museum Tusculanum Press, 1997), 94 (and n.307), 253; J. Weinstein, "The Chronology of Palestine in the Early Second Millennium B.C.E.," *BASOR* 288 (1992): 30–31.

13. Bietak, "Hyksos Rule," 109, 111. For Nehesy's family and possible Nubian ancestry see Ryholt, *Egypt*, 253–54. Although Ryholt believes Nehesy's father was Sheshi, other evidence in D. Ben-Tor, S. Allen, and J. Allen, "Seals and Kings," *BASOR* 315 (1999): 58–59, 61, argues against this. No text or inscription gives Nehesy's father's name, and though the Turin Papyrus gives Nehesy a one-year reign, he may have exercised authority as "King's Son" earlier.

14. Bietak, *Avaris and Piramesse*, 254–55; Dever, "Relations between Syria-Palestine and Egypt," 78.

15. Bietak, *Avaris and Piramesse*, 247; Holladay, "Nile Delta," 185–86; Dever, "Relations between Syria-Palestine and Egypt," 78; S. Manning, *A Test of Time: The Volcano of Thera and the Chronology and History of the Aegean and East Mediterranean in the Mid Second Millennium BC* (Oxford: Oxbow, 1999), 328.

16. Waddell, *Manetho*, 97, gives the total of 103 years for the Fifteenth Dynasty, but the Turin papyrus is closer to the time in question and in general is a good deal more accurate.

17. Ryholt, *Egypt*, 366–76.

18. Redford, *Egypt, Canaan, and Israel*, 110–11; A. Kempinski, "Some Observations on the Hyksos (XVth) Dynasty and Its Canaanite Origins," in *Pharaonic Egypt: The Bible and Christianity*, ed. S. Israelit-Groll (Jerusalem: Magnes Press, 1985), 131–33; idem., "Jacob in History," *BAR* 14 (January–February 1988): 44–45; P. McCarter, "The Patriarchal Age: Abraham, Isaac and Jacob," in *Ancient Israel: A Short History from Abraham to the Roman Destruction of the Temple*, ed. H. Shanks (Washington: Biblical Archaeology Society, 1988), 11.

19. Kempinski, "Jacob in History," 45, 46. For mention of the redating, see, Ben-Tor, Allen, and Allen, "Seals and Kings," 69 (note 22).

20. Redford, *Egypt, Canaan, and Israel*, 125–29.

21. W. Petrie, *Hyksos and Israelite Cities*, British School of Archaeology in Egypt Pub. 12 (London, 1906), 28–34.

22. C. Redmount, *On an Egyptian/Asiatic Frontier: An Archaeological History of the Wadi Tumilat*. (Chicago: University of Chicago Department of Near Eastern Languages and Civilizations Ph.D. Dissertation, 1989), 20; G. Sestini, "Nile Delta: A Review of Depositional Environments and Geological History," in *Deltas: Sites and Traps for Fossil Fuels*, ed. M. Whateley and K. Pickering, Geological Society Spec. Pub. 41 (London, 1989), 121.

23. M. Bietak, *Tell el-Dab'a II. Der Fundort im Rahmen einer archäologisch-geographischen Untersuchung über das ägyptische Ostdelta*, Denkschriften der Gesamtakademie 4 (Vienna: Österreichische Akademie der Wissenschaften, 1975), 88–90.

24. Redmount, *Egyptian/Asiatic Frontier*, 56–57.

25. Redmount, *Egyptian/Asiatic Frontier*, 12.

26. Redford, "Exodus I 11," 403–405; Redmount, *Egyptian/Asiatic Frontier*, 14.

27. Redford, "Exodus I 11," 405–408.

28. Redmount, *Egyptian/Asiatic Frontier*, 72.

29. Redmount, *Egyptian/Asiatic Frontier*, 177.

30. Holladay, "Nile Delta," 223–26.

31. C. Redmount, "Pots and Peoples in the Egyptian Delta: Tell El-Maskhuta and the Hyksos," *Journal of Mediterranean Archaeology* 8, no. 2 (1995): 67; Holladay, "Nile Delta," 194, 195–96.

32. Redmount, *Egyptian/Asiatic Frontier*, 257–59.

33. Holladay, "Nile Delta," 195.

34. Redmount, *Egyptian/Asiatic Frontier*, 252–56; Holladay, "Nile Delta," 188.

35. The cooking pots appear in stratum H at Tell el-Dab'a and in the equivalent to stratum E/1 at Tell el-Maskhuta. These strata are separated by at least 150 years in Bietak's chronology and by several hundred years by William G. Dever: idem., "Settlement Patterns and Chronology of Palestine in the Middle Bronze Age," *Hyksos*, 295.

36. Redmount, *Egyptian/Asiatic Frontier*, 244, 791–98, and idem., "Pots," 68, 71, 74.

37. Redmount, *Egyptian/Asiatic Frontier*, 231, 239, 265.

38. Redmount, *Egyptian/Asiatic Frontier*, 252, 264 (quote), and idem., "Pots," 78, 82–83.

39. Holladay, "Nile Delta," 203–204, 208–209.

40. Holladay, "Nile Delta," 208.

41. Redmount, *Egyptian/Asiatic Frontier*, 57.

42. Redmount, *Egyptian/Asiatic Frontier*, 256, 265, 903.

43. Weinstein, "Chronology of Palestine," 32–33, and idem., "Reflections on the Chronology of Tell el-Dab'a," in *Egypt, the Aegean and the Levant: Interconnections in the Second Millennium* B.C., ed. W. Davies and L. Schofield (London: British Museum, 1995), 87.

44. Redmount, *Egyptian/Asiatic Frontier*, 265, gives her own and Weinstein's first estimated dates; Weinstein, "Reflections," 88, gives his later estimated date range.

45. Holladay, "Nile Delta," 188.

46. Weinstein, "Reflections," 85; Bietak, "Hyksos Rule," 90.

47. W. Ward and W. Dever, *Scarab Typology and Archaeological Context: An Essay on Middle Bronze Age Chronology* (San Antonio: Van Siclen, 1994), 61.

48. W. Dever, "The Chronology of Syria-Palestine in the Second Millennium B.C.E.: A Review of Current Issues," *BASOR* 288 (1992): 1–25; idem., "Relations between Syria-Palestine and Egypt," 75, 78.

49. For the first radiocarbon date see Bietak, *Avaris and Piramesse*, 233. For the other two see F. Hassan and S. Robinson, "High-Precision Radiocarbon Chronometry of Ancient Egypt, and Comparisons with Nubia, Palestine, and Mesopotamia," *Antiquity* 61 (1987): 124, 133.

50. Manning, *Test of Time*, 326–28; S. Manning et al., "Chronology for the Aegean Late Bronze Age 1700–1400 B.C.," *Science* 312 (2006): 568 (fig. 3).

51. Holladay, "Nile Delta," 188, 204; Bietak, "Hyksos Rule," 90.

52. Redmount, *Egyptian/Asiatic Frontier*, 72.

53. Holladay, "Nile Delta," 204.

54. J. Bourriau, "Relations between Egypt and Kerma during the Middle and New Kingdoms," in *Egypt and Africa: Nubia from Prehistory to Islam*, ed. W. Davies (London: British Museum, 1991), 130.

55. R. Said, *The River Nile: Geology, Hydrology, and Utilization* (New York: Pergamon, 1994), 132; D-J. Stanley et al., "Nile Flow Failure at the End of the Old Kingdom, Egypt: Strontium Isotopic and Petrologic Evidence," *Geoarchaeology* 18 (2003): 399 (fig. 2).

56. Holladay, "Nile Delta," 204.

57. Manning's dates and the discussion, and his assignment of the Minoan eruption of Santorini to the D/3 stratum at Avaris are found in *Test of Time*, 326–28.

Chapter 3 The Minoan Eruption

1. P. Betancourt and G. Weinstein, "Carbon-14 and the beginning of the Late Bronze Age in the Aegean," *American Journal of Archaeology* 80 (1976): 329–48; H. Michael, "Radiocarbon Dates from Akrotiri on Thera," *Temple University Aegean Symposium, First*, P. Betancourt, ed. (Philadelphia: Temple University Press, 1976), 7–9.

2. For a discussion of the dating controversy up to 1999: Manning, *Test of Time*. Analysis of volcanic glass particles from the 1645 ± 4 B.C.E. layer in the Greenlandic

GRIP ice core shows they came from Alaska's Aniakchak volcano: N. Pearce et al., "Identification of Aniakchak (Alaska) tephra in Greenland ice core challenges the 1645 BC date for Minoan eruption of Santorini," *Geochemistry Geophysics Geosystems* 5 (2004): Q03005, doi:10.1029/2003GC000672. The growth spurt in the Anatolian tree ring chronology (Manning, *Test of Time*, 307–18; S. Manning et al., "Anatolian Tree-Rings and a New Chronology for the East Mediterranean Bronze-Iron Ages," *Science* 294 (2001): 2532–35), is likely from Aniakchak's eruption, since a major high-latitude eruption (such as Aniakchak) is thought to produce high-latitude stratospheric heating and a correspondingly negative North Atlantic Oscillation (NAO): A. Robock, "Volcanic Eruptions and Climate," *Reviews of Geophysics* 38 (2000): 209. A negative NAO has been linked to increased precipitation over Anatolia: H. Cullen and P. deMenocal, "North Atlantic Influence on Tigris-Euphrates Streamflow," *International Journal of Climatology* 20 (2000): 853–63. This increased precipitation would have temporarily increased growth in Anatolian trees. S. Manning et al., "Confirmation of Near-Absolute Dating of East Mediterranean Bronze-Iron Dendrochronology," *Antiquity* 77 (2003): http://antiquity.ac.uk/ProjGall/Manning/manning.html tightly constrained the onset of this growth spurt by radiocarbon dating to 1653–1651 B.C.E. (+7/−3 years). This is six to eight years older than the ice core date of the Aniakchak eruption, a fact which *may* signal a shortfall in the ice core record of six to eight years in this period.

3. Manfred Bietak recently wrote that the date of the Thera eruption was not settled, with the dating on historical grounds "in 1500 B.C.E. or even slightly later:" M. Bietak, "The Volcano Explains Everything—Or Does It?" *BAR* 32 (November–December 2006), 63.

4. Manning, *Test of Time*, 328–29.

5. For the dates on carbonized seeds from Thera: S. Manning et al., "Chronology for the Aegean Late Bronze Age 1700–1400 B.C.," *Science* 312 (2006): 566. For the discovery and dating of the olive branch: W. Friedrich et al., "Santorini Eruption Radiocarbon Dated to 1627–1600 B.C.," *Science* 312 (2006): 548. Friedrich et al.'s table S2 shows that with a 25% error the 95.4% probability extends from 1635 to 1591 B.C.E., and with a 50% error the 95.4% probability range is 1654–1597 B.C.E.

6. For the correlation of the GRIP and Dye 3 cores and their acid spikes, see H. Clausen et al., "A Comparison of the Volcanic Records over the Past 4000 Years from the Greenland Ice Core Project and Dye 3 Greenland Ice Cores," *JGR* 102 (1997), 26,707–26,723. The Dye 3 core, in southern Greenland, receives greater precipitation than the GRIP core; thus the Dye 3 core is less likely to be missing any annual snow deposits and so its year count is preferred. Conversely, melting and refreezing in the Dye 3 layers mean that some volcanic signals do not show up as strongly (or for as long) there as they do in the GRIP core. For example, the 1622 B.C.E. VAD (volcanic acid deposited) signal in the Dye 3 core lasted for only half a year, is only a 10 and rated at G1 (the lowest volcanic acid signal rating, with G4 being the highest), while the corresponding 1618 B.C.E. VAD signal in the GRIP core is 37 and G3 and lasted for 1.3 years (see Clausen et al., tables 2 and 3). Comparison of ^{10}Be concentrations in the GISP2 and GRIP ice cores with ^{14}C concentrations from tree ring chronologies shows that the GISP2 time scale is stretched by 50–60 years after 3300 or 3400 years B.P.: J. Southon, "A First Step to Reconciling the GRIP and GISP2 Ice-Core Chronologies, 0–14,500 yr B.P.,"

Quaternary Research 57 (2002): 32–37. Southon notes that the 3644 B.P. GISP2 sulfur peak correlates with the Dye 3-3593 B.P. and the GRIP-3585 B.P. acid peaks. Giving preference to the Dye 3 year count produces an offset of 51 years for the GISP2 core for this period. Thus the 1669 B.C.E. spike in the GISP2 core (G. Zielinski et al., "Record of Volcanism Since 7000 B.C. from the GISP2 Greenland Ice Core and Implications for the Volcano-Climate System," *Science* 264 [1994], table 2) would actually be at 1618 B.C.E. The sulfuric acid in the GISP2-1669 B.C.E. (corrected to1618 B.C.E.) acid spike (78 parts per billion) is above the level thought to mark the Minoan eruption (46 parts per billion): Manning, *Test of Time*, 277 (fig. 53).

7. Manning et al., "Near-Absolute Dating."

8. For the 1628 B.C.E. tree ring anomaly: Manning, *Test of Time*, 264–65.

9. A. Scarth, *Volcanoes: An Introduction* (College Station: Texas A & M University Press, 1994), 4, 23–24; W. Friedrich, *Fire in the Sea* (Cambridge: Cambridge University Press, 2000), 22–25.

10. D. Pyle, "New Estimates for the Volume of the Minoan Eruption," *Thera III-2*, 113–21; H. Sigurdsson, S. Carey, and J. Devine, "Assessment of Mass, Dynamics and Environmental Effects of the Minoan Eruption of Santorini Volcano," *Thera III-2*, 103, 104; H. Sigurdsson et al., "Marine Investigations of Greece's Santorini Volcanic Field," *Eos* 87 (August 22, 2006): 337, 342.

11. C. Doumas, *Thera: Pompeii of the Ancient Aegean* (London: Thames and Hudson, 1983), 134–38; G. Heiken and F. McCoy, "Precursory Activity to the Minoan Eruption, Thera, Greece," *Thera III-2*, 79–88.

12. For the BO₁ phase and its deposits on Thera: F. McCoy and G. Heiken, "The Late-Bronze Age Explosive Eruption of Thera (Santorini), Greece: Regional and Local Effects," in *Volcanic Hazards and Disasters in Human Antiquity*, ed. F. McCoy and G. Heiken, Geological Society of America Spec. Paper 345 (Boulder, 2000), 49–50: R. Sparks and C. Wilson, "The Minoan Deposits: A Review of Their Characteristics and Interpretation," *Thera III-2*, 90. For the plinian deposits in four deep sea cores southeast of Santorini: R. Sparks and T. Huang, "The Volcanological Significance of Deep-Sea Ash Layers Associated with Ignimbrites," *Geological Magazine* 117 (1980): 426–28, 433.

13. A. Scarth, *Vulcan's Fury: Man against the Volcano* (New Haven: Yale University Press, 1999), 153–54; S. Winchester, *Krakatoa The Day the World Exploded: August 27, 1883* (New York: Harper Collins, 2003), 295–96.

14. F. McCoy and G. Heiken, "Tsunami Generated by the Late Bronze Age Eruption of Thera (Santorini) Greece," *Pure and Applied Geophysics* 157 (2000): 1236–37.

15. McCoy and Heiken, "Eruption of Thera," 50–52; A. Scarth and J.-C. Tanguy, *Volcanoes of Europe* (Oxford: Oxford University Press, 2001), 84; Scarth, *Vulcan's Fury*, 31–33, 37–38.

16. McCoy and Heiken, "Tsunami," 1237–39; Scarth, *Vulcan's Fury*, 154. For the evidence from northeastern Crete: E. Hadingham, "Did a Tsunami Wipe Out the Cradle of Western Civilization?" *Discovery* 29 (January 2008), 8–14.

17. McCoy and Heiken, "Eruption of Thera," 52–54; Sparks and Wilson, "Minoan Deposits," 91–93; McCoy and Heiken, "Tsunami," 1239.

18. A. Woods and K. Wohletz, "Dimensions and Dynamics of Co-ignimbrite Eruption Columns," *Nature* 350 (1991): 225–27; Pyle, "New Estimates," 114–18; McCoy and

Heiken, "Tsunami," 1238; R. Sparks et al., "Sedimentology of the Minoan Deep-Sea Tephra Layer in the Aegean and Eastern Mediterranean," *Marine Geology* 54 (1983): 141–54. For the tephra deposits from Gölcük and Black Sea: F. Guichard et al., "Tephra from the Minoan Eruption of Santorini in Sediments of the Black Sea," *Nature* 363 (1993): 610–12.

19. G. Heiken and F. McCoy, "Caldera Development during the Minoan Eruption, Thira, Cyclades, Greece," *JGR* 89 (1984): 8441–62. Walter Friedrich (*Fire in the Sea*, 76–77), however, believes these are reworked deposits and not a separate eruption phase.

20. K. Kastens and M. Cita, "Tsunami-Induced Sediment Transport in the Abyssal Mediterranean Sea," *GSABull* 92 (1981): 845–57; M. Cita et al., "New Findings of Bronze Age Homogenites in the Ionian Sea: Geodynamic Implications for the Mediterranean," *Marine Geology* 55 (1984): 47–62; McCoy and Heiken, "Tsunami," 1240–41.

21. E. Pielou, *The Energy of Nature* (Chicago: University of Chicago Press, 2001), 78–81; E. Prager, *Furious Earth: The Science and Nature of Earthquakes, Volcanoes, and Tsunamis* (New York: McGraw-Hill, 2000), 176–77.

22. McCoy and Heiken, "Tsunami," 1243–46; K. Minoura et al., "Discovery of Minoan Tsunami Deposits," *Geology* 28 (2000): 59–62; D. Neev, N. Bakler, and K. Emery, *Mediterranean Coasts of Israel and Sinai* (New York: Taylor & Francis, 1987), 56–57. Tel Michal is also in the path of the southeast-directed tsunami(s) evidenced from Palaikastro in Crete: Hadingham, "Tsunami," 8–14.

23. W. Rose, Jr., G. Bluth, and G. Ernst, "Integrating Retrievals of Volcanic Cloud Characteristics from Satellite Remote Sensors: A Summary," *Philosophical Transactions of the Royal Society (London)* 358A (2000): 1593.

24. Blong, *Time of Darkness*, 64.

25. L. Kittleman, "Geologic Methods in Studies of Quaternary Tephra," in *Volcanic Activity and Human Ecology*, ed. P. Sheets and D. Grayson (New York: Academic Press, 1979), 57.

26. Guichard et al., "Tephra," 610, 611.

27. See McCoy and Heiken, "Eruption of Thera," figure 13 and references in the caption of figure 3.1 for sites where Minoan tephra was recovered.

28. D.-J. Stanley and H. Sheng, "Volcanic Shards from Santorini (Upper Minoan Ash) in the Nile Delta, Egypt," *Nature* 320 (1986): 734. The major element compositions of the Nile Delta glass shards are somewhat atypical when compared to other Minoan glass, but this may reflect analytical uncertainties and/or alteration. P. Bitschene and H.-U. Schmincke, "Fallout Tephra Layers: Composition and Significance," in *Sediments and Environmental Geochemistry: Selected Aspects and Case Histories*, ed. D. Heling et al.(Berlin: Springer,1990), 51, and J. Hunt and P. Hill, "Tephrological Implications of Beam Size–Sample-Size Effects in Electron Microprobe Analysis of Glass Shards," *Journal of Quaternary Science* 16 (2001): 105–17 discuss these analytical problems. Nile Delta glass refractive indices range from 1.507 to 1.510, well within the Minoan range (1.509 ± 0.003): J. Keller et al., "Explosive Volcanic Activity in the Mediterranean over the Past 200,000 yr as Recorded in Deep-Sea Sediments *GSABull* 89 (1978): 595. The only other glassy Mediterranean tephra with similar indices of refraction, Avellino (1.511 ± 0.003), is dissimilar in its chemical composition to the

Nile Delta glass and to any tephra from Santorini: B. Narcisi and L. Vezzoli, "Quaternary Stratigraphy of Distal Tephra Layers in the Mediterranean—an Overview," *Global and Planetary Change* 21 (1999): table 1. Sizewise, the Nile Delta glass particles, ranging from 20 to 60 microns (Stanley and Sheng, 734), are smaller than the wind-borne Theran glass particles from Anatolian lakes (>150 microns): N. Pearce et al., "Trace-element composition of single glass shards in distal Minoan tephra from SW Turkey," *Journal of the Geological Society of London* 159 (2002): 545–56, and also smaller than wind-borne Minoan particles from Black Sea cores, which range up to 106 microns: Guichard et al., "Tephra," 611 (table 1). Thus the Nile Delta glass was likely wind-borne also.

29. D. Morrison, "Ancient Egypt Cities Leveled by Massive Volcano, Lava Find Suggests," *National Geographic News* (April 2, 2007): http://news.nationalgeographic.com/news/pf/36949218.html.

30. G. Rumney, *Climatology and the World's Climates* (New York: MacMillian, 1968), 248; Scarth and Tanguy, *Volcanoes of Europe*, 84.

31. D. Sullivan, "The Discovery of Santorini Minoan Tephra in Western Turkey," *Nature* 333 (1988): 554.

32. McCoy and Heiken, "Eruption of Thera," 58.

33. Rumney, *Climatology*, 252; J. Lockwood, *World Climatology: An Environmental Approach* (New York: St. Martin's, 1974), 115–16.

34. Sestini, "Nile Delta," 102.

35. A. Bond and R. Sparks, "The Minoan Eruption of Santorini, Greece," *Journal of the Geological Society of London* 132 (1976): 9, 15.

36. S. de Silva and J. Alzueta, "The Socioeconomic Consequences of the A.D. 1600 Eruption of Huaynaputina, Southern Peru," in *Volcanic Hazards and Disasters in Human Antiquity*, ed. F. McCoy and G. Heiken, Geological Society of America Spec. Paper 345 (Boulder, Co., 2000), 17–18; K. Briffa et al., "Influence of Volcanic Eruptions on Northern Hemisphere Summer Temperature over the Past 600 Years," *Nature* 393 (1998): 452–53; J. Cole-Dai, E. Mosley-Thompson, and L. Thompson, "Annually Resolved Southern Hemisphere Volcanic History from Two Antarctic Ice Cores," *JGR* 102 (1997): 16765, 16768.

37. S. de Silva and G. Zielinski, "Global Influence of the AD 1600 Eruption of Huaynaputina, Peru," *Nature* 393 (1998): 455–58; de Silva and Alzueta, "Socioeconomic Consequences," 15–24.

38. Scarth, *Vulcan's Fury*, 142–45; Winchester, *Krakatoa*, 232–40;

39. Vitaliano, *Legends of the Earth*, 188–89.

40. N. Watkins et al., "Volume and Extent of the Minoan Tephra from Santorini Volcano: New Evidence from Deep-Sea Sediment Cores," *Nature* 271 (1978): 122–26.

41. W. Rose, Jr., et al., "Ice in the 1994 Rabaul Eruption Cloud: Implications for Volcano Hazard and Atmospheric Effects," *Nature* 375 (1995): 477–79.

42. N. Oskarsson, "The Interaction between Volcanic Gases and Tephra: Fluorine Adhering to Tephra of the 1970 Hekla Eruption," *JVGR* 8 (1980): 250–66; A. Tabazedeh and R. Turco, "Stratospheric Chlorine Injection by Volcanic Eruption: HCL Scavenging and Implications for Ozone," *Science* 260 (1993): 1082–86; Rose et al., "Rabaul Eruption," 477–79; Rose, Bluth, and Ernst, "Integrating Retrievals," 1590, 1595, 1597–98;

M. Herzog et al., "The Effect of Phase Changes of Water on the Development of Volcanic Plumes," *JVGR* 87 (1998): 56–57.

43. W. Rose, Jr., et al., "Observations of Volcanic Clouds in Their First Few Days of Atmospheric Residence: The 1992 Eruptions of Crater Peak, Mount Spurr Volcano, Alaska," *Journal of Geology* 109 (2001): 677–95; W. Rose, Jr., C. Riley, and S. Dartevelle, "Sizes and Shapes of 10-Ma Distal Fall Pyroclasts in the Ogallala Group, Nebraska," *Journal of Geology* 111 (2003): 115–24.

44. McCoy and Heiken, "Eruption of Thera," 48. An estimated $1.8–2.4 \times 10^{10}$ kilograms of chlorine and $1.8–2.7 \times 10^{11}$ kg of sulfur were released into the atmosphere during the Minoan eruption: V. Michaud, R. Clocchiatti, and S. Sbrana, "The Minoan and post-Minoan Eruptions, Santorini (Greece), in the Light of Melt Inclusions: Chlorine and Sulphur Behavior," *JVGR* 99 (2000): 213.

45. Nile Delta volcanic glass typically has large surface areas (Stanley and Sheng, "Nile Delta," fig. 3). High surface areas and aspect ratios will enhance atmospheric drag, and so the glass will remain aloft longer: see Rose, Riley, and Dartevelle, "Ogallala Group," 118–20.

46. R. Blong, *Volcanic Hazards* (Sydney: Academic Press, 1984), 66.

47. Blong, *Hazards*, 124, 329; P. Bauer, Review of Hekla. A Notorious Volcano by Sigurdur Thorarinsson, *Science* 172 (1971): 692–93.

48. P. Baxter et al., "Mount St. Helens Eruptions: The Acute Respiratory Effects of Volcanic Ash in a North American Community," *Archives of Environmental Health* 38, no. 3 (1983): 138–43: J. Nania and T. Bruha, "In the Wake of Mount St Helens," *Annals of Emergency Medicine* 11, no. 4 (1982): 184–91.

49. Blong, *Hazards*, 332.

50. G. Martin, "The Recent Eruption of Katmai Volcano in Alaska," *National Geographic* 24, no. 2 (1913): 180. Reindeer on the Alaskan Peninsula were also badly affected by the eruption of Unimak Volcano in 1825 and 1826: W. Workman, "The Significance of Volcanism in the Prehistory of Subarctic Northwest North America," in *Volcanic Activity and Human Ecology*, ed. P. Sheets and D. Grayson (New York: Academic Press, 1979), 353.

51. R. Wilcox, "Some Effects of Recent Volcanic Ash Falls with Especial Reference to Alaska," U.S. Geol. Survey Bulletin 1028-N (Washington, D.C.: Government Printing Office), 454; Blong, *Hazards*, 336–38.

52. L. A. Faustino, "Mayon Volcano and Its Eruption," *Philippine Journal of Science* 40, no. 1 (1929): 39.

53. Blong, *Hazards*, 326–27.

54. Blong, *Hazards*, 141–43, has studies of the mental health effects of volcanic eruptions and other studies involving natural disasters in general.

55. Scarth, *Vulcan's Fury*, 216–19, 223.

56. P. Sheets, "Environmental and Cultural Effects of the Ilopango Eruption in Central America," in *Volcanic Activity and Human Ecology*, ed. P. Sheets and D. Grayson (New York: Academic Press, 1979), 548–49; I. Burton, R. Kates, and G. White, *The Environment as Hazard*, 2nd ed. (New York: Guildford, 1993), 57–58, 121, 220–24; Scarth, *Vulcan's Fury*, 66–68, 193–94, 206.

57. de Silva and Alzueta, "Socioeconomic Consequences," 17; Blong, *Hazards*, 129; S. Coleman, *Volcanoes New and Old* (New York: John Day, 1946), 73.

58. J. Watson, "Krakatoa's Echo?" *Journal of the Polynesian Society* 72 (1963): 152 (quote); Blong, *Time of Darkness*, 71; idem., "Time of Darkness Legends from Papua New Guinea," *Oral History* 7, no. 10 (1979): 8, 21, 25.

59. Burton, Kates, and White, *Environment*, 57, 224–28.

60. For the Athapascans: D. Moodie, A. Catchpole, and K. Abel, "Northern Athapascan Oral Traditions and the White River Volcano," *Ethnohistory* 39 (1992): 148–71. For the El Salvador migration: Sheets, "Ilopango," 548–53.

61. Blong, *Hazards*, 142. Many publications describe disaster and counter-disaster syndromes, such as: National Institutes of Mental Health, 1978, *Training Manual for Human Service Workers in Major Disasters* (U.S. Dept. of Health, Education and Welfare Pub. No. (ADM) 77–538).

62. For Hesiod and his description of the Minoan eruption: M. T. Greene, *Natural Knowledge in Preclassical Antiquity* (Baltimore: Johns Hopkins University Press, 1992), 46–72 (especially the table on 60–63); Barber and Barber, *When They Severed Earth from Sky*, 82–85. For the Hittite myth: idem., 86–87. Unfortunately no ending to this text has survived, and so we do not know if it once described how the ash, once severed from its base, fell on the land of the Hittites, as ash from the co-ignimbrite eruption column would have done.

63. Barber and Barber, *When They Severed Earth from Sky*, 50.

64. H. Goedicke, "The Canaanite Illness," in *Festschrift Wolfgang Helk*, Studien zur altägyptischen Kultur 11, ed. H. Altenmüller and D. Wildung (Hamburg: Helmut Buske Verlag, 1984), 94, 98–99.

65. Bietak, "Hyksos Rule," 115 and n.63; Ryholt, *Egypt*, 308–09 n. 1077; Redford, "Textual Sources," 18.

Chapter 4 The Plagues, the Exodus, and Historical Reality

1. R. Friedman, *Who Wrote the Bible?* (New York: Summit, 1987), 24–27.

2. J. Van Seters, *The Life of Moses: The Yahwist as Historian in Exodus-Numbers* (Louisville: Westminster/John Knox, 1994), 78, 79.

3. M. Noth, *Exodus: A Commentary* (Philadelphia: Westminster, 1962), 70–71.

4. G. Fohrer, *Überlieferung und Geschichte des Exodus: Eine Analyse von Ex 1–15*, Beihefte zur Zeitschrift für die Alttestamentliche Wissenschaft 91 (Berlin: Alfred Töpelmann, 1964): 70; B. Childs, *The Book of Exodus: A Critical, Theological Commentary* (Philadelphia: Westminster, 1974), 131.

5. M. Greenberg, "The Thematic Unity of Exodus iii–xi," in *Fourth World Congress on Jewish Studies*, vol. 1 (Jerusalem: World Union of Jewish Studies, 1967), 53.

6. D. McCarthy, "Moses' Dealings with Pharaoh: Ex 7,8–10, 27," *Catholic Biblical Quarterly* 27 (1965): 336–45; G. Coats, *Moses: Heroic Man, Man of God* (Sheffield: JSOT Press, 1988), 103–05.

7. R. de Vaux, *The Early History of Israel* (Philadelphia: Westminster, 1978), 363.

8. Redmount, *Egyptian/Asiatic Frontier*, 228, 264.

9. Redmount, "Pots," 77–78.

10. For planting times along the Nile: B. Fagan, *Floods, Famines, and Emperors, El Niño and the Fate of Civilizations* (New York: Basic Books, 1999), 104. For tending donkey caravanners: Holladay, "Nile Delta, 195, 203–04, 208–09.

11. Hort, "Plagues of Egypt," 49; J. Wilkinson, *A Popular Account of the Ancient Egyptians*, Vol. 2 (New York: Bonanza Books, 1989), 21–22.

12. Ian Wilson calculated (*Exodus: The True Story*, 113) that the eruption plume would have to be forty-eight kilometers high to be visible in the Delta.

13. For tsunami heights in open sea: McCoy and Heiken, "Tsunami," 1248 (fig. 12), 1249. Aegean tsunamis will reach northeastern Egypt in 45 or 50 minutes: A. Yalciner et al., "Modeling and Visualization of Tsunamis: Mediterranean Examples," in *Tsunami and Nonlinear Waves*, ed. A. Kundu (Berlin: Springer, 2007), 278–81.

14. Sestini, "Nile Delta," 102.

15. Blong, *Time of Darkness*, 166; B. Bolt, *Earthquakes* (New York: Freeman, 1988), 85.

16. D. Irvin, "The Joseph and Moses Stories as Narrative in the Light of Ancient Near Eastern Narrative," *IJH*, 199.

17. For the text: M. Lichtheim, *Ancient Egyptian Literature*, Vol.1, *The Old and Middle Kingdoms* (Berkeley: University of California Press, 1973), 151. For its Hyksos date: J. Van Seters, "The Date for the 'Admonitions' in the Second Intermediate Period," *JEA* 50 (1964): 13–23.

18. F. Round, *The Ecology of Algae* (Cambridge: Cambridge University Press, 1981), 307.

19. J. Lenes et al., "Iron Fertilization and the *Trichodesmium* Response on the West Florida Shelf," *Limnology and Oceanography* 46 (2001), 1261–77.

20. *Trichodesmium*, unlike most phytoplankton, may utilize both dissolved and particulate iron (Lenes et al., "*Trichodesmium* Response," 1261–62). Even more important, acidification of dust by sulfuric acid (such as found in volcanic emissions) will cause the dust-borne iron to be readily utilized by phytoplankton: M. Meskhidze, W. Chameides, and A. Nenes, "Dust and Pollution: A Recipe for Enhanced Ocean Fertilization," *JGR* 110 (2005): D03301.

21. Herzog et al., "Phase Changes," 55–56.

22. R. Cook et al., "Impact on Agriculture of the Mount St. Helens Eruptions," *Science* 211 (1981): 18.

23. Blong, *Time of Darkness*, 113.

24. All biblical quotations (fewer than 500 verses are quoted) are from the New Revised Standard Version (Oxford and New York: Oxford University Press). Copyright 1989 by the Division of Christian Education of the National Council of the Churches of Christ in the United States of America. (MT) refers to verse numbers in the Masoritic text when verse citations differ from those in the New Revised Standard Version.

25. Martin, "Eruption of Katmai," 131–81; Workman, "Significance of Volcanism," 339–71.

26. J. Hoffmeier, *Israel in Egypt: The Evidence for the Authenticity of the Exodus Tradition* (New York: Oxford, 1997), 147.

27. Rose, Riley, and Dartevelle, "Ogallala Group," 115–24.

28. Hort, "Plagues of Egypt," 50.

29. Blong, *Time of Darkness*, 142–43, fig. 53.

30. Burton, Kates, and White, *Environment as Hazard*, 121–23; Sheets, "Ilopango," 548–49.

31. Ryholt, *Egypt*, 307.

32. E. Oren, "The 'Kingdom of Sharuhen' and the Hyksos Kingdom," *Hyksos*, 266.

33. Sheets, "Ilopango," 549.

34. de Vaux, *Israel*, 370.

35. de Vaux, *Israel*, 373; G. Coates, "Despoiling the Egyptians," *Vetus Testamentum* 18 (1968), 451.

36. Noth, *Exodus*, 53–54; Hoffmeier, *Israel in Egypt*, 87.

37. A. Cabrol, "Les mouflons du dieu Amon-Rê," in *Egyptian Religion: The Last Thousand Years*, ed. W. Clarysse, A. Schoors, and H. Willems (Leuven: Uitgeverij Peeters and Dept. Oosterse Studies, 1998), 532–33, 537–38. Mass slaughter of sheep or goats such as found in the Passover was not characteristic of Egyptian sacrifice: S. Mercer, *The Religion of Ancient Egypt* (London: Luzac, 1949), 355–56; S. Sauneron, *The Priests of Ancient Egypt* (Ithaca, N.Y.: Cornell University Press, 2000), 160–61.

38. The remains of sheep and other animal sacrifices were found in the Hyksos strata at Tell el-Dab'a (Bietak, *Avaris and Piramesse*, 250–51, 257) and in Middle Bronze Age levels in southern Canaan such as at Tel Haror: Oren, "Sharuhen," 264.

39. Hort, "Plagues of Egypt," 54–55.

40. Van Seters, *Moses*, 119.

41. M. Lehner, "The Giza Plateau Mapping Project: 1993–94 Annual Report," www-oi.uchicago.edu/OR/AR/93-94/93-94_Giza.html; J. Grüss, "*Saccharomyces winlocki*, die Hefe aus den Pharaonengräbern," *Tageszeitung für Brauerei* 27 (1929): 275–78; idem., "Untersuchung von Broten aus der ägyptischen Sammlung der staatlichen Museen zu Berlin," *Zeitschrift für ägyptische Sprache und Altertumskunde* 68 (1932): 79–80; B. Kemp, *Ancient Egypt: Anatomy of a Civilization* (London: Routledge, 1989), 120–25, 151–54, 222–23, 289–90, describes Middle and New Kingdom methods of bread-making in Egypt and Nubia.

42. The tomb of Tuthmosis III's vizier Rekhmire includes paintings of bricklayers "captured by His Majesty," who are quoted as saying "he supplies us with bread and beer. . . ." See J. Finegan, *Light from the Ancient Past: The Archeological Background of the Hebrew-Christian Religion* (Princeton: Princeton University Press, 1951), 90–91. In addition, at Amarna in the reigns of Amenophis III and IV and at Memphis during the reign of Seti I (father of Ramesses II) flour was strictly controlled by royal or temple bureaucracies and bread was baked in large central bakeries, probably associated with temples, and distributed to workers as part of their food ration: Kemp, *Ancient Egypt*, 222–23, 289–90.

43. E. Fox, *Genesis and Exodus: A New English Rendition with Commentary and Notes* (New York: Schocken, 1990), 299.

44. For *km wr*: Hoffmeier, *Israel in Egypt*, 212. The land between Lake Timsah and the Bitter Lakes may have held an extension of a known late seventh or early sixth century B.C.E. frontier canal (Neco's Canal), that existed *north* of Lake Timsah, between Ismailia and Qantara. If this canal extended south of Lake Timsah it would "link Lake Timsah and the Bitter Lakes to the Pelusiac branch at Qantara. . . .": A. Shafei Bey,

"Historical Notes on the Pelusiac Branch, the Red Sea Canal and the Route of the Exodus," *Bulletin de la Société Royale de Géographie d'Égypte* 21 (1946): 242. With its present gradient, however, water from Lake Timsah, the Bitter Lakes and the Red Sea would flow back into the Nile and make its waters unfit for drinking (Shafei Bey, 244). This canal was possibly a redug Middle or New Kingdom canal that formed a border canal, pictured in the relief of Seti I: A. Sneh, T. Weissbrod, and I. Perath, "Evidence for an Ancient Egyptian Frontier Canal," *American Scientist* 63 (1975), 546, 548. Aristotle, in *Meterologia* 1.14:22–31 (H. Lee, translator, *Aristotle Meterologica* (Cambridge: Harvard University Press, 1952), 117) says that a king named Sesostris attempted to build a canal linking the Nile to the Red Sea, but did not complete it because the Red Sea was higher than the land, and seawater would flood into the fresh water of the Nile. Three kings named Sesostris reigned in the Middle Kingdom's Twelfth Dynasty. The high (16 or possibly 18.5 meters) ridge called el Gisr north of Lake Timsah would have presented a formidable obstacle to these early canal builders (Shafei Bey, 244), although it might have been a good place for donkey caravans to cross over from the Sinai Peninsula into the Wadi Tumilat (see chapter 2), in which case it was probably guarded by a fort.

45. The corridor between the two Bitter Lakes was a likely place for the Exodus because the water there was shallow enough to ford: J. Simons, *Geographical and Topographical Texts of the Old Testament* (Leiden: Brill, 1959), 248, 250–51; M. Har-el, *The Sinai Journeys: The Route of the Exodus* (San Diego: Ridgefield, 1983), 313, 351, 353. For the north Red Sea ridge, see N. Voltzinger and A. Androsov, "Modeling the Hydrodynamic Situation of the Exodus," *Izvestiya, Atmospheric and Oceanic Physics* 39 (2003): 482–96, especially figs. 1 (483), 8 (492), and 9 (493).

46. Vitaliano, *Legends of the Earth*, 264 suggested that an east wind would have driven the water in the Great Bitter Lake basin to its northwestern end, exposing the land ridge between the two lakes, while the water in the Little Bitter Lake would have piled against its western shore, not against the land ridge. Such an east wind could have come on the northern edge of the cyclonic storm system as it moved southeastward from the Mediterranean. Voltzinger and Androsov "Modeling," found that a wind with a speed of 28–32 meters per second blowing for eight to ten hours causes the water over the Gulf of Suez ridge, normally under 4 meters deep at low tide, to shallow to a depth of 20 to 25 centimeters and to stay shallow for about four hours after the wind stops blowing. Data from a Red Sea drill core reported in M. Siddall et al., "Sea-Level Fluctuations during the Last Glacial Cycle," *Nature* 423 (2003): 854 (fig. 1), show that the Red Sea was approximately two meters lower at about 1600 B.C.E. than it is today. It is thus quite probable that the water depth modeled by Voltzinger and Androsov is too high and the northern Gulf of Suez ridge would have been completely dry during and for some hours after a strong windstorm.

Chapter 5 Moses and the Mountain of God

1. Childs, *Exodus*, 19; de Vaux, *Israel*, 329.
2. N. Na'aman, "Habiru and Hebrews: The Transfer of a Social Term to the Literary Sphere," *JNES* 45 (1986): 272–75.

3. For Jacob's location in north-central Canaan: McCarter, "Patriarchal Age," 12, 16, 22, 24. For the famine: Bell, "Climate," 261; Ryholt, *Egypt*, 301, 304. Ryholt also notes evidence for recurrent famines during the earlier part of the Second Intermediate Period (late Thirteenth and Fourteenth Dynasties, Sixteenth Dynasty) in the north, central, and southern sections of Egypt. For Memphis and the list of Asiatics: Redford, *Egypt, Canaan and Israel*, 78; W. Albright, "Northwest-Semitic Names in a List of Egyptian Slaves from the Eighteenth Century B.C.," *Journal of the American Oriental Society* 74 (1954): 223, 229.

4. Artapanus is quoted in Eusebius, *Praeparatio Evangelica* 9.27:1–37. See J. Collins, "Artapanus (Third to Second Century B.C.): A New Translation and Introduction," in *The Old Testament Pseudepigrapha*, Vol. 2, ed. J. Charlesworth (Garden City: Doubleday, 1985), 889–903.

5. Artapanus also says that Palmanothes was the son of Mempsasthenoth and that Palmanothes forced the Jews to build Heliopolis and Tanis, and Chenephres was the pharaoh of the Exodus. Artapanus also recounts several other incidents in Moses' life, including some that are not found in the Bible. Looking at this narrative as a whole, both Collins and Ben Zion Wacholder, *Eupolemus: A Study of Judaeo-Greek Literature* (Cincinnati and New York: Hebrew Union College–Jewish Institute of Religion, 1974), 105, consider it to be fictitious. However, traditions are formed from accounts that first have grown shorter and become single anecdotes ("leveling"—see the Appendix) and then are aggregated around noted individuals or culture heroes: Vansina, *Tradition*, 21, 106. Artapanus' account should be evaluated by looking at its individual pieces, keeping in mind that fusion plays a key role in traditional sources and that more reliable items can be (and often are) combined or fused with less reliable and completely unrelated ones to form a narrative whole. Sequential reordering often takes place and items can be put into parts of a narrative where they don't belong (Vansina, 172). For example, Artapanus says that "Pharethothes" was the name of the Egyptian king when Abraham came to Egypt (*Praeparatio Evangelica* 9.18:1). This name is derived from Pharaoh and Thoth (Collins, 897, note c). The pharaoh Tuthmosis III (which means "the god Thoth is born") plays a key role in the second exodus (see Chapter 10).

6. For Chenephres being Khaneferre Sobekhotep IV: Waddell, *Manetho*, 73 (note 3). Archaeological evidence indicates that the Thirteenth Dynasty lost direct contact with Nubia after Neferhotep I, older brother of Sobekhotep IV: (1) after this time the Egyptian pottery in Nubia is purely Upper Egyptian, suggesting creation of a separate political and/or economic entity in Upper Egypt: Bourriau, "Egypt and Kerma," 130; (2) the cooking pots and food remains at the Egyptian Middle Kingdom second cataract fort of Askut show a dramatic increase in Nubian influence in the mid- to late Thirteenth Dynasty: S. Smith, *Wretched Kush: Ethnic Identities and Boundaries in Egypt's Nubian Empire* (London: Routledge, 2003), 118–24, table 5.6. For Khaneferre Sobekhotep IV's wives: Ryholt, *Egypt*, 230.

7. Redford, *Egypt, Canaan, and Israel*, 78.

8. Ben-Tor, Allen, and Allen, "Seals and Kings," 67 (note 5).

9. T. Meek, "Moses and the Levites," *American Journal of Semitic Languages and Literatures* 56 (1939): 118–19; de Vaux, *Israel*, 329.

10. Ryholt, *Egypt*, 300; H. Smith, "The story of Onchsehshonqy," *Serapis: The American Journal of Egyptology* 6 (1980): 154.

11. R. Wilson, "Genealogy, Genealogies," in *Anchor Bible Dictonary*, Vol. 2 (New York: Doubleday, 1992), 930.

12. B. Childs, "The Birth of Moses," *JBL* 84 (1965): 109–22; de Vaux, *Israel*, 328.

13. For the Sargon birth tale see H. Saggs, *Babylonians* (Norman: University of Oklahoma Press, 1995), 66–67.

14. Miller, "Introduction," 16.

15. Redmount, *Egyptian/Asiatic Frontier*, 903; Holladay, "Nile Delta," 224.

16. F. Winnett, "The Arabian Genealogies in the Book of Genesis," in *Translating and Understanding the Old Testament*, ed. H. Frank and W. Reed. (Nashville: Abingdon, 1970), 188, 191–92.

17. W. Albright, "Midianite Donkey Caravans," in *Translating and Understanding the Old Testament*, ed. H. Frank and W. Reed. (Nashville: Abingdon, 1970), 197–205.

18. G. Hort, "Musil, Madian, and the Mountain of the Law," in *Jewish Studies: Essays in Honour of the Very Reverend Dr. Gustav Sicher, Chief Rabbi of Prague* (Prague: Council of Jewish Religious Communities, 1955), 81–93.

19. D. Almond, "Geological Evolution of the Afro-Arabian Dome," *Tectonophysics* 131 (1986): 321–22, 327 (fig. 7d); H.-J. Bayer et al., "Sedimentary and Structural Evolution of the Northwest Arabian Red Sea Margin," *Tectonophysics* 153 (1988): 147; V. Camp and M. Roobol, "The Arabian Continental Alkali Basalt Province: Part I. Evolution of Harrat Rahat, Kingdom of Saudi Arabia," *GSABull* 101 (1989): 93–94; idem., "Upwelling Asthenosphere beneath Western Arabia and Its Regional Implications," *JGR* 97 (1992): 15255–71.

20. Camp and Roobol, "Basalt Province," 71; D. Almond, "The Relation of Mesozoic-Cainozoic Volcanism to Tectonics in the Afro-Arabian Dome," *JVGR* 28 (1986): 229, 238–39.

21. G. Brown, D. Schmidt, and A. Huffman, Jr., *Geology of the Arabian Peninsula: Shield Area of Western Saudi Arabia*, U.S. Geological Survey Prof. Paper 560-A (Washington: Government Printing Office, 1989), A154. Note that Harrat 'Uwayrid lavas are alkali-olivine basalts falling within the hawaiite field on the plagioclase-color index normative diagram.

22. Francis, *Volcanoes*, 121.

23. R. Linton, "Nativistic Movements," *American Anthropologist* 45 (1943): 230–40; A. Wallace, "Revitalization Movements," *American Anthropologist* 58 (1956): 264–81; for the quote: M. Wise, *The First Messiah* (San Francisco: HarperSanFrancisco, 1999), 2.

24. Tupi-Guarani prophetic migrations: R. Ribeiro, "Brazilian Messianic Movements," in *Millennial Dreams in Action*, ed. S. Thrupp (The Hague: Mouton, 1962), 55–69; H. Clastres, *The Land-Without-Evil: Tupí-Guaraní Prophetism* (Urbana: University of Illinois Press, 1995). Peter the Hermit and similar prophet-leaders of the First Crusade: H. Lamb, *The Crusades* (New York: Bantam, 1962), 55–64.

25. Flavius Josephus, *Antiquities of the Jews* 2:264, 3:76; Flavius Josephus, *Against Apion* 2:25.

26. de Vaux, *Israel*, 435.

27. J. Kœnig, "La Localisation du Sinai et les Traditions des Scribes," *Revue d'Histoire et de Philosophie Religieuses* 44 (1964): 200–35 gives a review of these early ideas. Also Noth, *Exodus*, 109.

28. M. Noth, "Der Wallfahrtsweg zum Sinai," *Palästinafahrbuch* 36 (1940), 5–28.

29. C. Doughty, *Travels in Arabia Deserta* (London: Jonathan Cape and Medici Society, 1923 ed.), 408, 418; H. Philby, *The Land of Midian* (London: Ernest Benn, 1957), 140–44.

30. A. Musil, *The Northern Heğâz* (New York: American Geographical Society, 1926), 214. Hala'-l-Bedr is actually on the northern side of Tadra.

31. A. Musil, "Vorbericht über seine letzie Reise nach Arabien," *Anzerger der kaiserlichen Akademie der Wissenschaften in Wein, Philosophisch-historische Klasse* 48 (1911): 139–59; idem., *Northern Heğâz*, 215, 269–72; R. Coleman, *Geologic Evolution of the Red Sea* (New York: Oxford University Press, 1993), 57.

32. H. von Wissmann, "3,1 The Volcanoes of West Arabia," in *Catalog of Active Volcanoes of the World Including Solfatara Fields: Part 16, "Arabia and the Indian Ocean,"* ed. N. Van Padang (Rome: IAVCEI, 1963), 1–2.

33. Brown, Schmidt, and Huffman, *Arabian Peninsula*, A153; Philby, *Midian*, 142.

34. Philby, *Midian*, 155, 188.

35. Henige, *Historiography*, 103.

36. I am indebted to one of the reviewers of this book, Zeilinga de Boer, for the suggestion of the magma rising through the oil-bearing rock layers, causing a naturally burning oil well, not uncommon in Arabia.

37. For the phases of the moon see http://sunearth.gsfc.nasa.gov/eclipse/phase/phases-1699.html.

38. "Shavuot," *Encyclopedia Judaica* Volume 14 (Jerusalem: Keter Publishing, 1971), 1319–22; M. Weinfeld, "Pentecost as Festival of the Giving of the Law," *Immanuel* 8, no. 1 (1978), 7–12.

39. F. Cross, Jr., *Canaanite Myth and Hebrew Epic* (Cambridge: Harvard University Press, 1973), 169.

40. G. Davies, *The Way of the Wilderness* (Cambridge: Cambridge University Press, 1979), 65. Yet Davies, like most biblical scholars, rejects an Arabian location for the Mountain of God.

41. Childs, *Exodus*, 344.

42. For the documentary sources and their accounts: B. Schwartz, "What Really Happened at Mount Sinai?" *Bible Review* 13 (October, 1997): 20–30, 46.

43. Childs, *Exodus*, 349, notes several problems with the documentary hypothesis and concludes that the different traditions were combined in the oral stage of the tradition. See also Coats, *Moses*, 131; Y. Avishur "The Narrative of the Revelation at Sinai (Ex 19–24)," in *Studies in Historical Geography and Biblical Historiography presented to Zecharia Kallai*, Vetus Testamentum Supplement 81 (Leiden: Brill, 2000), 197–214.

44. B. Feiler, *Walking the Bible: A Journey by Land through the Five Books of Moses* (San Francisco: HarperCollins, 2001), 251, 269, describes the difficulty of the route to the summit of Jebel Musa, even using the steps that were cut over 1500 years after Moses.

45. H. Von Wissmann, "Volcanoes of West Arabia," 4; G. Brown et al., "Geologic Map of the Northwestern Hijaz Quadrangle, Kingdom of Saudi Arabia," U.S. Geological Survey Miscellaneous Geologic Investigations Map I-204 A (Washington: U.S. Geological Survey, 1963). Scale 1:500,000.

46. C. Graesser, "Standing Stones in Ancient Palestine," *Biblical Archaeologist* 35 (1972): 37; U. Avner, "Sacred Stones in the Desert," *BAR* 27 (May–June 2001): 32–33. Davies, "Wilderness Itineraries,"174, notes that at Har Karkom in the Negev and near Wadi Arabah are groups of twelve standing stones.

47. J. Kœnig, *Le Site de Al-Jaw dans l'Ancien Pays de Madian* (Paris: Geuthner, 1971).

48. J. Pirenne, "Le Site Préislamique de Al-Jaw, la Bible, le Coran et le Midrash," *Révue Biblique* 82 (1975): 34–69; Doughty, *Arabia Deserta*, 405.

49. Brown et al., "Geologic Map," shows the granite outcrops closest to Hala'-l-Bedr (as well as all the other types of rocks).

50. Pirenne's critique is considerably weakened by her acceptance of Doughty's outdated and long disproved idea that a volcano (Hala'-l-Bedr) could eject plutonic rocks such as granite (Pirenne, 37). She also states that the cinder cone was too small to produce a column of smoke that could be seen for long distances; however this eruption could also involve magma moving through oil-rich sandstones. Pirenne presents a far more convincing case for the liklihood that Kœnig designated the wrong cinder cone as Hala'-l-Bedr (and thus analyzed the wrong pile of blocks), and for thinking that the shapes of these rocks are the products of natural erosion.

51. Brown, Schmidt, and Huffman, *Arabian Peninsula*, A153–A154 note that where dated Nabatean inscriptions are exposed to the wind near the ground, sandblasting and spalling has undercut as much as a meter into the sandstone in 2,000 years.

52. J. Kirsch, *Moses: A Life* (New York: Ballantine, 1998), 279.

53. J. Kœnig, "Itinéraires Sinaïtiques en Arabie," *Révue de l'Histoire des Religions* 166 (1964), 137.

54. H. Moldenke and A. Moldenke, *Plants of the Bible* (Waltham: Chronica Botanica, 1952), 23–25, discuss the name "seneh" and the various species of acacia, including *Acacia tortilis* and *Acacia seyal* (*A. ehrenbergiana?*), candidates for the Hebrew name *seneh* (see next note).

55. Brown, Schmidt, and Huffman, *Arabian Peninsula*, A11, note that "Acacia, or camel thorn, is the most widespread, and in many places almost the only, shrub and small tree of the western desert. The acacia "samr" (*Acacia tortilis* [Forsk] Hayne), perhaps the hardiest, may be found growing along the wadis or, where rainfall is slightly higher, on the desert plains as isolated flat-topped trees or copses." When rainfall reaches four inches annually the sallam acacia (*A. ehrenbergiana*) becomes more common (see also A13–A15). Where rainfall is more abundant these species grow larger than bush size, and may have provided the wood for the ark.

56. Davies, *Wilderness*, 51.

57. Musil, *Northern Heğâz*, 365 (Retame) and references to this citation on this page.

58. Van Seters, *Moses*, 287.

59. Kœnig, "Localisation du Sinai," 202, note 4.

60. Davies, *Wilderness*, 59–60.

61. Noth, "Wallfahrtsweg zum Sinai," 5–28.

62. Kœnig, "Localisation du Sinai," 201–03.

63. Davies, "Wilderness Itineraries," 170. In rough terms, the eleven days between Kadesh and Mount Harb would fit the thirty kilometer per day distance

64. Musil, *Northern Heğâz*, 321–31; Hort, "Musil, Madian," 83.

Chapter 6 The Sojourn in the Wilderness

1. Oren, "Sharuhen," 257; D. Ilan, "The Dawn of Internationalism—The Middle Bronze Age," *ASHL*, 311–12; A. Mazar, *Archaeology of the Land of the Bible—10,000–586 B.C.E.* (New York: Doubleday, 1992), 198–208.

2. Oren, "Sharuhen," 255; Ilan, "Internationalism," 302–06.

3. Oren, "Sharuhen," 271–73.

4. Ilan, "Internationalism," 306.; Oren, "Sharuhen," 273.

5. L. Cassan, *The Ancient Mariners*, 2nd ed. (Princeton: Princeton University Press, 1991), 10.

6. J. Gray, *Archaeology and the Old Testament World* (New York: Harper & Row, 1962), 88; N. Na'aman, "The Hurrians and the End of the Middle Bronze Age in Palestine," *Levant* 26 (1994): 176–81.

7. Gray, *Archaeology*, 88–89.

8. M. Anbar and N. Na'aman, "An Account Tablet of Sheep from Ancient Hebron," *Tel Aviv* 13–14 (1986–1987), 3–12.

9. W. Dever, "Palestine in the Second Millennium BCE: The Archaeological Picture," *IJH*, 85–86.

10. Weinstein, "Reflections," 88.

11. Oren, "Sharuhen," 257; Ilan, "Internationalism," 311–12; Mazar, *Archaeology*, 198–208.

12. This number is found in Exodus 38:26, Numbers 1:46 and 2:32.

13. Friedman, *Wrote*, 192–96 says that the writer of "P" intentionally transformed the story of the rebellion of the sons of Reuben into a rebellion by Korah and the Levites. Oral tradition, however, may already have been at work, fusing these two stories.

14. G. Hort, "The Death of Qorah," *Australian Biblical Review* 7 (1959): 20–26.

15. Vansina, *Tradition*, 132–33.

16. P. J. King, "Circumcision—Who Did it, Who Didn't and Why," *BAR* 12 (July–August 2006): 48–55.

17. See Exodus 20:5; 34:7, Numbers 14:18, and Deuteronomy 5:9.

18. These verses are: 24:1–2, 9–11, 13*a*, 14–15*a*; 34:10–13, 14*b*–16; 32:17–18, 25–29: de Vaux, *Israel*, 400.

19. S. Herrmann, *A History of Israel in Old Testament Times* (Philadelphia: Fortress, 1981), 74.

20. Friedman, *Wrote*, 190.

21. Scholars who believe Aaron to be a later addition include: G. Coats, *Rebellion in the Wilderness* (Nashville: Abingdon, 1968), 67; Noth, *Exodus*, 51; Friedman, *Wrote*, 188–206. Friedman, however, believes the writer of "P" intentionally made Aaron the equal of Moses to uphold the role and power of the Aaronid priesthood.

22. For olivine phenocrysts in the harrat basalts: Brown, Schmidt, and Huffman, *Arabian Peninsula*, A154; Doughty, *Arabia Deserta*, 405.

23. Hort, "Qorah," 13–14.

24. 1 Kings 18:38.

25. For the genealogies of Caleb: 1 Chronicles 2:9; Numbers 32:12; Joshua 14:6.

26. de Vaux, *Israel*, 537.

27. See Exodus 31:1–11; 37; 38:1–22 for the making of the ark and its furnishings.

28. Later many of the Kenites settled with or alongside the Jerahmeelites in the Negev (1 Samuel 30:29), while some of them settled down with the Amalakites (1 Samuel 15:5–6).

29. R. Hendrix, "A Literary Structural Analysis of the Golden-Calf Episode in Ex 32:1–33:6," *Andrews University Seminary Studies* (1990): 211–17; C. Hayes, "Golden Calf Stories: The Relationship of Exodus 32 and Deuteronomy 9–10," in *The Idea of Biblical Interpretation: Essays in Honor of James L. Kugel*, ed. H. Nazman and J. Newman (Leiden: Brill, 2004), 45–93. As Hayes compellingly demonstrates, Deuteronomy 9–10 is the first midrash on Exodus 32, rather than being its source, and the similarity between the Exodus 32 story and the passage in 1 Kings 12:26–32 where King Jeroboam makes two golden calves and places them in Shechem and Dan (see M. Aberbach and L. Smolar, "Aaron, Jeroboam and the Golden Calves," *JBL* 86 [1967], 129–40) is best seen as an ironic satire directed against Jeroboam for his cultic worship.

30. Numbers 20:14–21 (a much later story) is sandwiched within the story of Aaron's disobedience and death because it too is set at Kadesh (Numbers 20:22–23 was added later to connect the two stories). Numbers 33:37–41*a* was also inserted into the itinerary at a relatively late date. The sequence in that section originally went: Kadesh and/or Wilderness of Zin to Zalmonah to Punon. In Deuteronomy 10:6, Aaron is said to have died at Moserah, a different locality altogether.

31. Numbers 27:14, Deuteronomy 32:51, and Ezekiel 48:28.

32. Noth, *Exodus*, 139, thinks the name Massah ("to test") is secondary to the story. Coats, *Rebellion*, 58, also regards the testing motif as secondary.

33. Hayes, "Golden Calf," 63–64.

34. In Deuteronomy 10:6–9 there is a remnant memory of the association of Aaron's death with the setting apart of the tribe of Levi to care for the ark, but this association has been transferred away from the mountain of God and to a later time. Hayes, "Golden Calf," 81–84, sees the association of Aaron's death with the dedication and responsibilities of the Levites as revisionist history, but such a use does not preclude its being a traditional remnant brought in for a revisionist purpose.

35. Exodus 32:19; Deuteronomy 9:17.

36. R. Thomas, *Literacy and Orality in Ancient Greece* (Cambridge: Cambridge University Press, 1992), 83–84, 86; Niditch, *Oral World*, 43, 55; Redford, *Egypt, Canaan, and Israel*, 87.

37. Doughty, *Arabia Deserta*, 420.

38. W. Dever, "Nelson Glueck and the Other Half of the Holy Land," in *The Archaeology of Jordan and Beyond*, ed. L. Stager, J. Greene, and M. Coogan (Winona Lake, Ind.: Eisenbrauns, 2000), 116.

39. I. Finkelstein, *Living on the Fringe* (Sheffield: Sheffield Academic Press, 1995), 23–30, 156; C.-M. Bennett, "Biblical Traditions and Archaeological Results," in *The Archaeology of Jordan and other Studies Presented to Siegfried H. Horn*, ed. L. Geraty and L. Herr (Berrien Springs: Andrews University Press, 1986), 80–81.

40. Ø. LaBianca and R. Younker, "The Kingdoms of Ammon, Moab, and Edom: The Archaeology of Society in Late Bronze Iron Age Transjordan (ca. 1400–500 BCE)," *ASHL*, 399–411.

41. Miller, "Introduction," 16; Vansina, *Tradition*, 24.

42. K. Kitchen, "The Egyptian Evidence on Ancient Jordan," in *Early Edom and Moab: The Beginning of the Iron Age in Southern Jordan*, ed. P. Bienkowski (Sheffield: Collis, 1992), 21–22.

43. Some modern scholars think that Kadesh is at the spring of 'Ain Qedeis, about fifty miles south-southwest of Beer-sheba in the Sinai Peninsula. However, archaeological excavations have turned up no occupation there earlier than the seventh century B.C.E.: R. Cohen, "Did I Excavate Kadesh-Barnea?" *BAR* 7 (May–June, 1981), 20–33.

44. J. M. Miller, "The Israelite Journey through (around) Moab and Moabite Toponymy," *JBL* 108 (1989): 581; Flavius Josephus, *Antiquities of the Jews* 4.82; Eusebius, *Onomasticon*, quoted in H. Bar-Deroma, "Kadesh-Barne'a," *Palestine Exploration Quarterly* 96 (1964): 107; Hort, "Qorah," 8. Davies, *Wilderness*, 11, 17–18 states (18): "There are so many reasons for putting Kadesh at Petra and none at all that we know for putting Mount Hor there, that it seems that Kadesh *must* have been located there first, and Mount Hor only subsequently!"

45. For Dhiban on Egyptian topographical lists: C. Krahmalkov, "Exodus Itinerary Confirmed by Egyptian Evidence," *BAR* 20 (September–October 1994): 57–58. For the rest of the itinerary, see J. M. Miller, "Journey," 579, 581, 582 (note 8).

Chapter 7 Meanwhile, Back in Civilization

1. M. Stuiver, P. Grootes, and T. Braziunas, "The GISP2 δ^{18}O Climate Record of the Past 16,500 Years and the Role of the Sun, Ocean, and Volcanoes," *Quaternary Research* 44 (1995): 345–46; A. Robock, "Volcanic Eruptions and Climate," *Reviews of Geophysics* 38 (2000), 191–219. In the winter after the 1991 eruption of Mount Pinatubo there were very large amounts of rainfall in Israel (and snow in Jerusalem). See C. Price et al., "A Possible Link between El Niño and Precipitation in Israel," *Geophysical Research Letters* 25 (1998): 3964, fig. 1.

2. Extensive tree growth is attested to in the Anatolian tree rings for this period: Manning, *Test of Time*, 308–11. T. Bryce, *The Kingdom of the Hittites* (Oxford: Clarendon Press, 1998), 72, suggests the first known Hittite king, Hattusili I, came to the throne about 1650 B.C.E. Na'aman, "Hurrians," 179–81, suggests that the campaigns of Hattusili I and his grandson, Mursili I, caused large-scale migrations in many parts of the Near East, including Canaan, beginning in the seventeenth century B.C.E.

3. F. Rothlisberger, *10 000 Jahre Gletschergeschichte der Erde* (Aarau, Frankfurt am Main and Salzburg: Aarau and Sauerländer, 1986), 70; J. Hass et al., "Synchronous Holocene Climatic Oscillations Recorded on the Swiss Plateau and at Timberline in the Alps," *The Holocene* 8 (1998): 301–09, gives a date range of 3,500–3,200 ^{14}C years B.P. for the Alpine cold phase. J. Grove, *The Little Ice Age* (London: Methuen, 1988), 303, cites a radiocarbon age of 3440 ± 60 B.P. (H-2913) for peat buried by glacial gravels at this time.

4. M. Magny, "Solar Influences on Holocene Climatic Changes Illustrated by Correlations between Past Lake-Level Fluctuations and the Atmospheric ^{14}C Record," *Quaternary Research* 40 (1993): 2–3. This glacial advance was 100–150 meters beyond the advance of the glaciers during the Little Ice Age: Grove, *Little Ice Age*, 301; M. Magny

"Successive Oceanic and Solar Forcing Indicated by Younger Dryas and Early Holocene Climatic Oscillations in the Jura," *Quaternary Research* 43 (1995): 279–85.

5. W. Bircher, "Dendrochronology Applied in Mountain Regions," in *Handbook of Holocene Palaeoecology and Palaeohydrology*, ed. B. Berglund (Chichester: Wiley-Interscience, 1986), 400–01. Warming in Europe began about 1200 B.C.E. and reached modern temperature levels about 1180 B.C.E.

6. J. Eddy, "Historical and Arboreal Evidence for a Changing Sun," in *The New Solar Physics*, ed. J. Eddy (Boulder, Co.: Westview, 1978), 11–33.

7. Magny, "Solar Influences," 3–5; J. Neumann, "Climatic Changes in Europe and the Near East in the Second Millennium BC," *Climatic Change* 23 (1993): 237 (quoting F. Schweingruber and coworkers at the Swiss Federal Institute for Forest, Snow and Landscape Research and using their figure as part of his figure 2).

8. F. Street-Perrott and R. Perrott, "Abrupt Climate Fluctuations in the Tropics: The Influence of Atlantic Ocean Circulation," *Nature* 343 (1990): 609; F. Gasse and E. Van Campo, "Abrupt Post-Glacial Climate Events in West Asia and North Africa Monsoon Domains," *Earth and Planetary Science Letters* 126 (1994): 442–51; R. Telford and H. Lamb, "Groundwater-Mediated Response to Holocene Climatic Change Recorded by the Diatom Stratigraphy of an Ethiopian Crater Lake," *Quaternary Research* 52 (1999): 73; F. Street-Perrott et al., "Drought and Dust Deposition in the West African Sahel: A 5500-Year Record from Kajemarum Oasis, Northeastern Nigeria," *The Holocene* 10 (2000): 296–97, 299.

9. J. Maley, "Middle to Late Holocene Changes in Tropical Africa and Other Continents: Paleomonsoon and Sea Surface Temperature Variations," *NATO*, 617–18, 625; F. Gasse and E. Van Campo, "Abrupt Post-Glacial Climate Events in West Asia and North Africa Monsoon Domains," 446, 447, 454; Telford and Lamb, "Groundwater-Mediated Response," 73; D. Jolly, R. Bonnefille, and M. Roux, "Numerical Interpretation of a High Resolution Holocene Pollen Record from Burundi," *Palaeogeography, Palaeoclimatology, Palaeoecology* 109 (1994): 362–63, 368. Nonarboreal pollen (indicating dryer conditions) increased at 3800 B.P. throughout East Africa and Ethiopia, correlating with lower East African lake levels. After 3200 B.P. (earlier fifteenth century B.C.E.), however, a new semideciduous forest formed.

10. Bell, "Climate," 243–44; F. Hassan, "Nile Floods and Political Disorder in Egypt," *NATO*, 6; R. Said, *River Nile*, 132–49. K. Butzer, "The Holocene Lake Plain of North Rudolph, East Africa," *Physical Geography* 1 (1980): 47, 48, 51–52, notes that Lake Rudolf seasonally overflowed into the White Nile drainage about 3250 ± 150 B.P. (in the Eighteenth Dynasty).

11. J. Stager, B. Cumming, and L. Meeker, "A High-Resolution 11,400-Yr Diatom Record from Lake Victoria, East Africa," *Quaternary Research* 47 (1997): 81–89.

12. J. Pickrell, "Aerial War Against Disease: Satellite Tracking of Epidemics is Soaring," *Science News* 161 (April 6, 2002), 218–19.

13. T. Butler, "Yersiniosis and Plague," in *Zoonoses: Biology, Clinical Practice and Public Health Control*, ed. S. Palmer, L. Soulsby, and D. Simpson (Oxford: Oxford University Press, 1998), 290.

14. D. Cavanaugh, "Specific Effect of Temperature upon Transmission of the Plague Bacillus by the Oriental Rat Flea, *Xenopsylla cheopis*," *American Journal of Tropical*

Medicine and Hygiene 20 (1971): 264–73; J. Poland, T. Quan, and A. Barnes, "Plague," in *Handbook of Zoonoses, Section A: Bacterial, Rickettsial, Chlamydial, and Mycotic*, ed. G. Beran, 2nd ed. (Boca Raton, Fla.: CRC Press, 1994), 98; S. Scott and C. Duncan, *Biology of Plagues: Evidence from Historical Populations* (Cambridge: Cambridge University Press, 2001): 58–59.

15. C. Gregg, *Plague: An Ancient Disease in the Twentieth Century*. Rev. ed. (Albuquerque: University of New Mexico Press, 1985), 106–11.

16. Butler, "Yersiniosis and Plague," 287–88; R. Perry and J. Fetherston, "*Yersinia pesti*—Etiologic Agent of Plague," *Clinical Microbiology Reviews* 10 (1997): 58; J.-N. Biraben, *Les Hommes et la Peste en France et dans les Pays Européens et Mediterranéens*, Vol. 1 (Paris: Mouton, 1975), 302–03.

17. R. D'Arrigo et al., "Spatial Response to Major Volcanic Events in or about AD 536, 934 and 1258: Frost Rings and Other Dendrochronological Evidence from Mongolia and Northern Siberia: Comment on R. Strothers, 'Volcanic Dry Fogs, Climate Cooling, and Plague Pandemics in Europe and the Middle East'" *Climatic Change* 49 (2001): 239–42.

18. For the East African origin of the sixth century plague see P. Sarris, "The Justinianic Plague: Origins and Effects,"*Continuity and Change* 17 (2002): 170–72. For the transport route and means see D. Keyes, *Catastrophe: An Investigation into the Origins of the Modern World* (New York: Ballantine, 1999), 24.

19. W. McNeill, *Plagues and Peoples* (Garden City: Doubleday, 1977), 142–47. Scott and Duncan, *Biology of Plagues*, 389, present evidence that the Black Death of the fourteenth and seventeenth centuries in England was not caused by *Yersinia pestis* but rather by a viral hemorrhagic fever such as Ebola. They suggest that two plagues, bubonic in Asia and hemorrhagic in Europe (especially northern Europe) coexisted from the fourteenth through the seventeenth centuries.

20. A. Appleby, "The Disappearance of Plague: A Continuing Puzzle," *Economic History Review* 33 (1980): 163–65.

21. S. Nicholson, "Saharan Climates in Historic Times," in *The Sahara and the Nile: Quaternary Environments and Prehistoric Occupation in Northern Africa*, ed. M. Williams and H. Faure (Rotterdam: Balkema, 1980), 178–83; R. Strothers, "Climate and Demographic Consequences of the Massive Volcanic Eruption of 1258," *Climatic Change* 45 (2000): 366–67.

22. R. Strothers, "Volcanic Dry Fogs, Climate Cooling, and Plague Pandemics in Europe and the Middle East," *Climatic Change* 42 (1999), 720.

23. Keyes, *Catastrophe*, 24.

24. Redford, *Egypt, Canaan, and Israel*, 119–20.

25. Redford, "Textual Sources," 17. This text is several hundred years after the events in question, and the main story is probably partly legendary: Redford, *Egypt, Canaan, and Israel*, 233. However, there is no reason to include a plague in the story unless it is a historic holdover.

26. For Hecataeus of Abdera see M. Stern, *Greek and Latin Authors on Jews and Judaism*. Vol. 1 (Jerusalem: Magnes Press, 1974), 20–44.

27. J. Harris and K. Weeks, *X-Raying the Pharaohs* (New York: Scribners, 1973), 122–23.

28. Redford, *Egypt, Canaan, and Israel*, 125–27; idem., "Textual Sources," 14–15.

29. Ryholt, *Egypt*, 278, 280; E. Wente, "Genealogy of the Royal Family," in *An X-Ray Atlas of the Royal Mummies*, ed. J. Harris and E. Wente (Chicago: University of Chicago Press, 1980), 124.

30. Redford, *Egypt, Canaan, and Israel*, 128.

31. Redford, "Textual Sources,"15.

32. Waddell, *Manetho*, 87–89.

33. B. Wood, "Did the Israelites Conquer Jericho?" *BAR* 16 (March-April 1990), 51.

34. H. Goedicke, "The End of the Hyksos in Egypt," in *Egyptological Studies in Honor of Richard A. Parker*, ed. L. Lesko (Hanover, N.H.: Brown University Press, 1986), 37; Redford, *Egypt, Canaan, and Israel*, 128.

35. Waddell, *Manetho*, 87–89; Redford, "Textual Sources," 15; Bietak, *Avaris: The Capital of the Hyksos* (London: British Museum Press, 1996), 67.

36. Flavius Josephus, *Against Apion* 2:21, 22, 23. William Whiston's older translation uses the word "buboes," while the more recent Thackeray translation in the Loeb Classical Library series uses the word "tumors." Since the word bubo refers specifically to a swelling of the lymphatic gland, especially in the groin, while the word tumor is a more general term for any sort of swelling, in this context bubo is clearly the preferred alternative: "they all had buboes (tumors) in their groins" (*Against Apion* 2:21).

37. Goedicke, "Canaanite Illness," 93–95, 103–04.

38. Na'aman, "Hurrians,"182–83; R. Giveon, *Les Bédouins Shosou des Documents Egyptiens* (Leiden: Brill, 1971), 9–10; J. Breasted, *Ancient Records of Egypt: Historical Documents*, Vol. 2, *The Eighteenth Dynasty* (Chicago: University of Chicago Press, 1906), 34–35, 50–51.

39. Gregg, *Plague*, 73. A black rat was found in the stomach of a mummy of an Egyptian sacred bird of about 1000 B.C.E. (Gregg, 65). An ulcer on the mummy of Rameses V (died about 1143 B.C.E.) may be a plague bubo: R. Hendrickson, *More Cunning Than Man: A Social History of Rats and Men* (New York: Stein and Day, 1983), 46–47.

40. Keyes, *Catastrophe*, 14–16.

41. K. Kenyon, *Excavations at Jericho*, Vol. 1, *The Tombs Excavated in 1952–4* (London: British School of Archaeology in Jerusalem, 1960), 443–504; idem., *Archaeology in the Holy Land*, 3rd ed. (New York: Praeger, 1970), 190–92; idem., *Digging Up Jericho* (London: E. Benn, 1957), 254–55. A seal with the possible prenomen of Khamudy, the last Hyksos ruler, was also found from a MB-LB Jericho tomb: Ryholt, *Egypt*, 52, 121, 387–88.

42. Definite archaeological evidence of the south Arabian spice trade goes back only to the Late Bronze Age, about 1300 B.C.E., when the domestication of the camel and "oasis urbanism" emerged in the Hejaz: P. Parr, "Edom and the Hejaz" in *Early Edom and Moab: The Beginning of the Iron Age in Southern Jordan*, ed. P. Bienkowski (Sheffield: Collis, 1992), 42, but trade is notoriously hard to detect archaeologically, and the widespread trade documented in the Middle Bronze Age suggests that spice trading via donkeys (rather than camels) probably started during that period: Oren, "Sharunen," 273; Holladay, "Nile Delta," 203–04. Handmade flat-bottomed cooking pots in Tell el-Maskhuta's earliest phases resemble later "so-called Midianite or Negebite wares, and

even to those cooking pots produced by recent Bedouin elements in the Wadi Tumilat" (Holladay, 190).

43. For the resemblance between the two Hebrew words and the translation as "pestilence": E. Neufeld "Insects as Warfare Agents in the Ancient Near East (Ex. 23:28; Deut. 7:20; Josh 24:12; Isa. 7:18–20)," *Orientalia* 49 (1980): 34–35, n. 8. Neufeld opts for the literal meaning of hornets and hypothesizes that they were used as agents of warfare, although he acknowledges that there is a total lack of cuneiform or pictorial evidence from Egypt or Mesopotamia of insects used in any military context (Neufeld, 39).

44. Numbers 25:1–2, 31:16.

45. W. Albright, "Donkey Caravans," 197–205; F. Cross, *Epic to Canon*, 62, n.30.

46. A. Christie, T. Chen, and S. Elberg, "Plague in Camels and Goats: Their Role in Human Epidemics," *Journal of Infectious Diseases* 141 (1980): 724–26.

47. R. Stieglitz, "Ancient Records and the Exodus Plagues," *BAR* 13 (November–December 1987), 49; R. Biggs, "Medicine, Surgery, and Public Health in Ancient Mesopotamia," in *Civilizations of the Ancient Near East*, Vol. 3, ed. J. Sasson (New York: Scribners, 1995), 1920.

Chapter 8 The Destruction of Jericho

1. E. Sellin and C. Watzinger, *Jericho: die Ergebnisse der Ausgrabungen* (Leipzig: Hinrichs, 1913).

2. C. Watzinger, "Zur Chronologie der Schichten von Jericho," *Zeitschrift der Deutschen Morgenländischen Gessellschaft* 80 (1926), 131–36.

3. For John Garstang's Jericho publications: Wood, "Israelites Conquer Jericho", 57 (notes 7–9). For Kathleen Kenyon's preliminary Jericho publications: Wood, "Israelites Conquer Jericho," note 16. For other Kenyon publications: Chapter 7, n. 41 and posthumous volumes edited by Thomas Holland: *Excavations at Jericho*, Vols. 3, 4, and 5 (London: British School of Archaeology in Jerusalem, 1981, 1982, 1983).

4. K. Kenyon, "Some Notes on the History of Jericho in the Second Millennium B.C.," *PEQ* 83 (1951), 101–38.

5. D. Neev and K. Emery, *The Destruction of Sodom, Gomorrah, and Jericho: Geological, Climatological, and Archaeological Background* (Oxford: Oxford University Press, 1995), 96.

6. For the structure and geologic history of the Dead Sea basin: Z. Garfunkel, and Z. Ben-Avraham, "The Structure of the Dead Sea Basin," *Tectonophysics* 266 (1996): 155–76.

7. For the Jericho fault: Neev and Emery, *Destruction*, 95, 96, 100. The ground on both sides of a normal fault is pulled, with one side moving downward relative to the other: Prager, *Furious Earth*, 51, 53 (fig. 2.7b).

8. A. Ben-Menahem, "Four Thousand Years of Seismicity along the Dead Sea Rift," *JGR* 96 (1991): 20,206.

9. K. Prag, "The Intermediate Early Bronze–Middle Bronze Age Sequences at Jericho and Tell Iktanu Reviewed,"*BASOR* 264 (1986): 63, 70–71.

10. D. Ussishkin, "Notes on the Fortifications of the Middle Bronze II Period at Jericho and Shechem," *BASOR* 276 (1989): 33; Kenyon, *Holy Land*, 178–79; Kenyon and Holland, *Excavations at Jericho*, Vol. 3, 375, plate 236.

11. A. Waltham, "Geological Hazards," in *The Cambridge Encyclopedia of Earth Sciences*, ed. D. Smith (Scarborough, Ontario: Prentice-Hall Canada and Cambridge University Press, 1981), 437 (see "Masonry D").

12. Kenyon estimated the Jericho destruction to be at the end of the Middle Bronze Age, about 1550 B.C.E.: K. Kenyon, "Jericho," in *Archaeology and Old Testament Study*, ed. D. Winton Thomas (Oxford: Clarendon, 1967), 272; idem., "Jericho" in *Encyclopedia of Archaeological Excavations in the Holy Land*, Vol. 2, ed. M. Avi-Yonah (Englewood Cliffs, N.J.: Prentice-Hall, 1976), 551, 554. For Jericho destruction layer radiocarbon dates: Manning, *Test of Time*, 257. This table uses a different computer program than that in Chapter 1 but includes the 1998 tree-ring data.

13. Kenyon, *Holy Land*, 197–98.

14. Wood, "Israelites Conquer Jericho," 51.

15. A. Nur, and E. Cline, "What Triggered the Collapse? Earthquake Storms," *Archaeology Odyssey* 4 (September–October 2001): 35.

16. Kenyon and Holland, *Excavations at Jericho,* Vol. 3, 370.

17. For a brief history of earthquake measurement: Prager, *Furious Earth*, 39–44, 57–62. For the Modified Mercalli Scale: E. Keller and N. Pinter, *Active Tectonics: Earthquakes, Uplift, and Landscape* (Upper Saddle River, N.J.: Prentice-Hall, 1996), 19.

18. Y. Enzel, G. Kadan, and Y. Eyal, "Holocene Earthquakes Inferred from a Fan-Delta Sequence in the Dead Sea Graben," *Quaternary Research* 53 (2000): 34. 46–47.

19. Ben-Menahem, "Four Thousand Years," 20,205; Enzel, Kadan, and Eyal, "Holocene Earthquakes," 34.

20. Neev and Emery, *Destruction*," 104.

21. Ben-Menahem, "Four Thousand Years," 20,205; de Vaux, *Israel*, 607.

22. Bruins and Van der Plicht, "Exodus Enigma," 213.

23. P. King and L. Stager, *Life in Biblical Israel* (Louisville: Westminster/John Knox, 2001), 94.

24. Keller and Pinter, *Active Tectonics*, 18, 20–21.

25. Keller and Pinter, *Active Tectonics*, 21–22 and fig. 1.12. Sellin and Watzinger found the revetment walls built upon 1.3 meters of hard clay, while later excavators stated they were built on bedrock. Because the bedrock is weathered and uneven, it's more likely that the revetment walls were built on an even layer of clay: Ussishkin, "Notes on the Fortifications," 32–33. Because they were bolstered front and rear, the revetment walls were protected from earthquakes.

26. Kenyon and Holland, *Excavations at Jericho*, Vol. 3, 110, plate 236.

27. Wood, "Israelites Conquer Jericho," 58 n. 47, credits William H. Shea with this suggestion.

28. R. Stieglitz, "Ancient Records," 49.

29. P. Bienkowski, *Jericho in the Late Bronze Age* (Warminster: Aris and Phillips, 1986), 112, 122.

Chapter 9 The Conquest and Settlement of Canaan

1. C. Herzog and M. Gichon, *Battles of the Bible* (London: Greenhill, 1997), 49.

2. A. Singer, *The Soils of Israel* (Berlin: Springer, 2007), 90–91.

3. Finkelstein, "Emergence of Israel," 173.

4. Herzog and Gichon, *Battles*, 51.

5. Z. Zevit, "Archaeological and Literary Stratigraphy in Joshua 7–8," *BASOR* 251 (1983): 26.

6. J. Callaway, "Excavating at Ai (Et-Tell): 1964–1972," *Biblical Archaeologist* 39, no. 1 (1976): 19.

7. A. Rainey, "Review of Bimson 1978," *IEJ* 30 (1980), 249–51; D. Livingston, Further Considerations on the Location of Bethel at El-Bireh," *PEQ* 126 (1994), 154–59.

8. Zevit, "Joshua 7–8," 25 (fig. 1), 29–30.

9. Zevit, "Joshua 7–8," 31.

10. Herzog and Gichon, *Battles*, 303 n. 7.

11. Herzog and Gichon, *Battles*, 51.

12. de Vaux, *Israel*, 631. J.Weinstein, "The Egyptian Empire in Palestine: A Reassessment," *BASOR* 241 (1981): 3, says that the evidence for a destruction level toward the end of MB IIC at Gibeon is not conclusive but that the site was not occupied in the Late Bronze Age I period. The Gibeonite towns may have been abandoned because of their proximity to Jerusalem.

13. Herzog and Gichon, *Battles*, 56.

14. A. Rainey, "The Biblical Shephelah of Judah," *BASOR* 251 (1983): 1–2, 6.

15. Judges 1:19, 21, 27–35 (see also Joshua 13:13; 15:63; 16:10; and 17:11–13, 16).

16. The later story is in Judges 4–5.

17. For Jabon of Qishon: Krahmallkov, "Exodus Itinerary Confirmed," 61–61 and 79 (notes 20 and 21). Krahmallkov also mentions an earlier Jabon, of Hazor, in the eighteenth-seventeenth centuries B.C.E. For Hazor's destruction: Y. Yadin, "Is the Biblical Account of the Israelite Conquest of Canaan Historically Reliable?" *BAR* 8 (March–April 1982): 19–20.

18. N. Na'aman, "The 'Conquest of Canaan' in the Book of Joshua and in History," *FNM*, 262.

19. Na'aman, "Conquest of Canaan," 262; de Vaux, *Israel*, 541–42, 637–38.

20. Na'aman, "Hurrians, 175; A. Zertal, "'To the Land of the Perizzites and the Giants': On the Israelite Settlement of the Hill country of Manasseh," *FNM*, 50, gives the figure of 135 Middle Bronze sites in Manasseh and 31 Late Bronze Age sites.

21. I. Finkelstein, *Settlement*, 341.

22. Weinstein, "Egyptian Empire," 1–28; idem. "Egypt and the Middle Bronze IIC/Late Bronze IA Transition in Palestine," *Levant* 23 (1991): 105–15.

23. A. Zertel, *The Israelite Settlement in the Hill Country of Manasseh* (Haifa: 1988), 219–20, quoted in S. Bunimovitz, "Socio-Political Transformations in the Central Hill Country in the Late Bronze–Iron I Transition," *FNM*, 193 n. 76.

24. J. Hoffmeier, "Reconsidering Egypt's Part in the Termination of the Middle Bronze Age in Palestine," *Levant* 21 (1989): 181–93.

25. P. Bienkowski, *Jericho in the Late Bronze Age* (Warminster: Aris & Phillips, 1986), 127–28.

26. Bunimovitz, "Central Hill Country," 181 and n. 10; Finkelstein, "Emergence of Israel," 173.

27. Finkelstein, *Settlement*, 342–43.

28. Na'aman, "Hurrians," 175–87; de Vaux, *Israel*, 87–89; Redford, *Egypt, Canaan, and Israel*, 137–40.

29. Neev, Bakler, and Emery, *Mediterranean Coasts of Israel and Sinai*, 2–3, 48, 56, 96, 100, 112, 114.

30. Neev, Bakler, and Emery, *Mediterranean Coasts of Israel and Sinai*, 3, 100–02.

31. J. Portugali, "Theoretical Speculations on the Transition from Nomadism to Monarchy," *FNM*, 209–11.

32. J. Morris, *The Age of Arthur: A History of the British Isles from 350 to 650* (New York: Scribner's, 1973), 222–24.

33. Na'aman, "Hurrians," 183.

34. L. Stager, "The Song of Deborah: Why Some Tribes Answered the Call and Others Did Not," *BAR* 15 (January–February 1989): 53.

35. R. Bookman et al., "Late Holocene Lake Levels of the Dead Sea," *GSABull* 116 (2005): 565; A. Frumkin, "The Holocene History of Dead Sea Levels," in *The Dead Sea: The Lake and Its Setting*, ed. T. Niemi, Z. Ben-Avraham, and J. Gat (New York: Oxford, 1997), 239 (fig. 22–4), 243; The two sets of estimates of the Dead Sea lake levels and their ages disagree because of the different evidence used: a rising salt diapir (Mount Sedom) in Frumkin's work and radiocarbon-dated lake sediments in Bookman et al.'s. Bookman et al. (p. 565) state that, for their Unit I (2140–1445 b.c.e.), they have no indicator for the absolute lake level elevation, leaving open the possibility that the lake was above 370 mbsl in the mid-sixteenth century b.c.e. Their Unit I corresponds to Frumkin's Stage 4 (2300–1500 b.c.e.). Frumkin (p. 243) states that the Dead Sea fell to a level of 380 MBLS *or lower* after the preceding Stage 3. This fall from an earlier highstand in the Dead Sea correlates with δO^{18} measurements on Galilee speleothems that signal a change from cooler to less cool (and probably dryer) conditions not long before 1500 b.c.e.: M. Bar-Matthews et al., "The Eastern Miditerranean Paleoclimate as a Reflection of Regional Events: Soreq Cave, Israel," *Earth and Planetary Science Letters* 166 (1999), 90 (fig. 2A).

36. Bookman, "Late Holocene Lake Levels," 565–66.

37. A. Frumkin and Y. Elitzur, "Historic Dead Sea Level Fluctuations Calibrated with Geological and Archaeological Evidence," *Quaternary Research* 57 (2002): 337, 340, and figs. 3 and 4.

38. Frumkin and Elitzur, "Dead Sea Level Fluctuations," 340, fig. 5.

39. Frumkin and Elitzur, "Dead Sea Level Fluctuations," 340–41 and fig. 5.

40. de Vaux, *Israel*, 641–42.

41. J. M. Miller, "The Israelite Occupation of Canaan," *IJH*, 229–30.

42. A. Zertal, "The Trek of the Tribes as they Settled in Canaan," *BAR* 17 (September–October 1991): 49; R. Frankel, "Upper Galilee in the Late Bronze–Iron I Transition," *FNM*, 22; Stager, "Song of Deborah," 62.

43. Pritchard, *Ancient Near Eastern Texts*, 245–47.

44. Zertal, "Trek of the Tribes," 49; Z. Gal, "Iron I in Lower Galilee and the Margins of the Jezreel Valley," *FNM*, 45.

45. Y. Aharoni (translated by A. F. Rainey), *The Land of the Bible: A Historical Geography* (Philadelphia: Westminster, 1979), 232–33; Stager, "Song of Deborah," 62–64.

46. Aharoni, *Land of the Bible*, 239.

47. Gal, "Lower Galilee," 38. For the term 'Apiru: Na'aman, "Habiru and Hebrews," 271–88.

48. de Vaux, *Israel*, 527–28, 533.

49. For the Mushite priesthood in the north and its rivalry with the Aaronite priesthood of Hebron: F. Cross, Jr., *Canaanite Myth and Hebrew Epic* (Cambridge: Harvard University Press, 1973), 195–209. For Levites in Ephraim: Judges 19:1, 18.

50. For the conquest of Hebron by Joshua and all Israel: Joshua 10:36–37. For its conquest by Caleb: Joshua 15:15.

51. For the conquest of Debir: Joshua 15:15–19 and Judges 1:13, 3:9.

52. J. Chadwick, "Discovering Hebron: The City of the Patriarchs Slowly Yields Its Secrets," *BAR* 31 (September–October 2005): 29–32.

53. A. Ofer, "'All the Hill Country of Judah': From a Settlement Fringe to a Prosperous Monarchy," 113; de Vaux, *Israel*, 535.

54. Judges 3:8–11.

55. de Vaux, *Israel*, 536–37 and n. 43.

56. Ofer, "Hill Country of Judah," 115 (table 1).

57. Numbers 32:33, 37–38; Joshua 13:15–23.

58. Cross, *From Epic to Canon*, 55–56 and n.8 (quote).

59. Cross, *From Epic to Canon*, 56 n. 8 and references therein.

60. de Vaux, *Israel*, 547–49; Ofer, "Hill Country of Judah," 112–16. For the formation of Judah during the reign of King David: Baruch Halpern, *David's Secret Demons: Messiah, Murderer, Traitor, King* (Grand Rapids: Eerdmans, 2001), 272, 296–97.

61. Finkelstein, *Settlement*, 200, 343–34.

62. Zertel, "Land of the Perizzites," 59.

63. Finkelstein, *Settlement*, 218–20, 343–44.

64. Na'aman, "Conquest of Canaan," 235.

65. S. Shennan, "Population, Culture History, and the Dynamics of Culture Change," *Current Anthropology* 41 (2000): 5–7 of the online edition; P. Pétrequin, "Management of Architectural Woods and Variations in Population Density in the Fourth and Third Millennia B.C. (Lakes Chalain and Clairvaux, Jura, France)," *Journal of Anthropological Archaeology* 15 (1996): 15–17; R. Pennington, "Causes of Early Human Population Growth," *American Journal of Physical Anthropology* 99 (1996): 271 (table 3).

66. J. Neel, "The Population Structure of an Amerindian Tribe, the Yamomana," *Annual Review of Genetics* 12 (1979): 367–68, 372. This increase occurred despite a female infanticide rate of 25% and the practice of late-term abortions.

67. Pennington, "Human Population Growth," 270–72 shows that the survival of young children was the most important factor in determining the rate of population increase and that "early child mortality rates are exquisitely sensitive to environmental factors (p. 264). P. Pétrequin, "Variations in Population Density," 16, clearly shows the link between population fluctuations and ^{14}C variations which signal climatic fluctuations.

68. L. Bruce-Chwatt, "Malaria in African Infants and Children in Southern Nigeria," *Annals of Tropical Medicine and Parasitology* 46 (1952): 190 (table 2), found that, in Lagos, Nigeria, malaria was responsible for 37.3% of all deaths of children aged three to four years and 18.6% in the five to ten year old group. The death rate for children was 12.5 per 1000 in the first two years, 6.45–6.75 per 1000 in ages two through four,

and 1.21 per thousand in ages five through ten years. In addition, exposure to malaria in the womb increases the rate of miscarriage and stillbirth, contributes to low birth weight, low infant weight in the first two years, and a higher death rate overall in children. G. Wernsdorfer and W. Wernsdorfer, "Social and Economic Aspects of Malaria and Its Control," in *Malaria: Principles and Practice of Malariology,* Vol. 2, ed. W. Wernsdorfer and I. McGregor (Edinburgh: Churchill Livingstone, 1988), 1424, report similar findings from other highly infectious areas, concluding that: "A considerable part of this high mortality in infants and young children is to be ascribed to malaria." As altitude increases and rainfall decreases, malaria becomes concentrated in progressively smaller valleys: J. Najera, B. Liese, and J. Hammer, *Malaria: New Patterns and Perspectives* (Washington: World Bank Technical Paper 183, 1992), 6.

69. Wernsdorfer and Wernsdorfer, "Aspects of Malaria," 1428: Pennington, "Human Population Growth," 263–64, 269, found lower mortality in young children was closely related to the availability of high-quality weaning foods, particularly milk from domesticated animals. Tribes with such animals experience low young-child mortality rates (Pennington, 263, table 1). Certain religious factors would also favor a higher Israelite fertility than that of their neighbors. Ritual prostitution practiced in certain Canaanite cults (Baal, for example) would increase the spread of pelvic inflammatory disease (caused by various sexually transmitted bacteria), which dramatically lowers fertility (see Pennington, 266, for pelvic inflammatory disease and fertility in Central Africa).

70. Na'aman, "Habiru and the Hebrews," 275–76; S. Bunimovitz, "On the Edge of Empires—Late Bronze Age (1500–1200 BCE)," *ASHL*, 327.

71. P. Newby, *Warrior Pharaohs: The Rise and Fall of the Egyptian Empire* (London: Faber & Faber, 1980), 116.

72. de Vaux, *Israel*, 639–40.

73. For the topographical list: Giveon, *Bédouins Shosou*, 26–28, 74–77.

74. Redford, *Egypt, Canaan, and Israel*, 272–73.

75. Ofer, "Hill Country of Judah," 96; only two of the site's six hectares were settled in the LBA.

Chapter 10 Back to Egypt

1. Na'aman, "Hurrians," 181–83.

2. In the tomb of his vizier Rekhmire, the length of Tuthmosis III's reign is reported as fifty-three years, ten months, and twenty-six days. For the royal family in the Eighteenth Dynasty: E. Wente, "Genealogy of the Royal Family," 122–56; idem., "Age at Death of Pharaohs of the New Kingdom Determined from Historical Sources," in *An X-Ray Atlas of the Royal Mummies*, ed. J. Harris and E. Wente (Chicago: University of Chicago Press, 1980), 238–86.

3. Redford, *Egypt, Canaan, and Israel*, 153–56; Giveon, *Bédouins Shosou*, 9–10; Newby, *Warrior Pharaohs*, 57–63.

4. Redford, *Egypt, Canaan, and Israel*, 156–58; Newby, *Warrior Pharaohs*, 63–67.

5. Newby, *Warrior Pharaohs*, 88.

6. Hoffmeier, *Israel in Egypt*, 113.

7. D. Redford, "An Egyptological Perspective on the Exodus Narrative," in *Egypt, Israel, Sinai: Archaeological and Historical Relationships in the Biblical Period*, ed. A. Rainey (Tel Aviv: Tel Aviv University Press, 1987), 144–45; Newby, *Warrior Pharaohs*, 82.

8. Hoffmeier, *Israel in Egypt*, 112, 114–15.

9. D. Benson, *Ancient Egypt's Warfare: A Survey of Armed Conflict in the Chronology of Ancient Egypt 1600 B.C.–30 B.C.* (Chicago: D. Benson, 1995), 63–64.

10. D. Redford, "The Coregency of Tuthmosis III and Amenophis II," *JEA*, 118–22. W. Murnane, in *Ancient Egyptian Coregencies*, Studies in Ancient Oriental Civilization 40 (Chicago: Oriental Institute, 1977), 44–48, argues that there was no military campaign into Syria in year three of Amenophis II's reign and that the two stelae at Amada and Elephantine which refer to this campaign, and are dated to year three and year four, respectively, were in fact erected in year eight and that *both* stelae fail to record the true year of the campaign, even though the stelae state they were set up by the explicit command of the monarch after his return from this campaign. This is an exceedingly unlikely argument, given that these monuments were made for posterity, not for the moment when "everybody at the time of the erection of the stelae knew that Amenophis II's first campaign had been fought in year seven" (Murnane, 48). For the day, month, and year of Tuthmosis III's death see W. Shea, "Exodus, Date of the," *International Standard Bible Encyclopedia*, Vol. 1 (Grand Rapids: Eerdmans, 1982), 234.

11. T. Säve-Söderbergh, *The Navy of the Eighteenth Egyptian Dynasty* (Uppsala: A.-B. Lundequistska, 1946), 34–38.

12. Friedman, *Wrote*, 143–44, believes that this verse was written by an exiled priest in Egypt (Dtr²) in the sixth century B.C.E. and that it refers to a remnant of Judeans fleeing Judah after the assassination of the Babylonian governor. However, he omits the words "in ships" ("the Lord will bring you back in ships to Egypt"). Fleeing Judeans could not have traveled by sea at that time because the Babylonians had destroyed all the coastal ports: E. Stern, *Archaeology of the Land of the Bible*, Vol. 2, *The Assyrian, Babylonian, and Persian Periods (732–332 B.C.E.)* (New York: Doubleday, 2001), 316–19. Furthermore, these Judeans were exiles, not slaves, as specified in the Deuteronomy text.

13. M. Bietak, *Avaris*, 70, 72, 81–82.

14. Hoffmeier, *Israel in Egypt*, 185; M. Abd El-Maksoud, *Tell Heboua (1981–1991): Enquête Archéologique sur la Deuxième Période Intermédiaire et le Nouvel Empire à l'exteémité Orientale du Delta* (Paris: Éditions Recherche sur les Civilisations, 1998), 36–37, 175, 177, 198, 200, 206. Level III has produced an abundance of Tuthmosis III pottery (Abd El-Maksoud, 37).

15. Newby, *Warrior Pharaohs*, 77–79 suggests that Tuthmosis III's dockyard was at Peru-nefer, but this name does not appear before the time of his son, Amenophis II: Redford, "Coregency," 109 n. 2. Amenophis II's stela at Giza refers to his base near Memphis as do other documents: M. Lichtheim, *Ancient Egyptian Literature*, Vol. 2, *The New Kingdom* (Berkeley, University of California Press, 1976), 42; Redford, "Coregency," 109. Also, men known to have lived in Peru-Nefer were buried at Sakkara, immediately north of Memphis: Säve-Söderbergh, *Navy*, 38.

16. For the Sinai forts of both the Eighteenth and Nineteenth Dynasties: E. D. Oren, "The 'Ways of Horus' in North Sinai," in *Egypt, Israel, Sinai: Archaeological and Historical Relationships in the Biblical Period*, ed. A. Rainey (Tel Aviv: Tel Aviv University Press,

1987), 69–119, especially 69–73. For the Nineteenth Dynasty defensive features: Sneh, Weissbrod, and Perath, "Ancient Egyptian Frontier Canal," 547–48.

17. J. Bennett, "Geophysics and Human History: New Light on Plato's Atlantis and the Exodus," *Systematics* 1 (1963): 150–54.

18. W. Shea, "Exodus, Date of the," 233–38.

19. Waddell, *Manetho*, 86, 101 (quote), 107–09 (quote), 111.

20. T. Higham, "The Experimental Study of the Transmission of Rumour," *British Journal of Psychology* 42 (1951), 51.

21. Israelite tradition may have originally remembered a first trip to Egypt by Abraham 400 (Genesis 15:13) or 430 (Exodus 12:40) years before the final exodus under Pharaoh Tuthmosis in about 1450 B.C.E. This date for Abraham's entry would be in rough agreement with the hypothesized birth of Jochebed, Abraham's great-great granddaughter, in about 1740 B.C.E. (see chapter 5), with 400 years (1850 B.C.E.) being the better fit. Using Ryholt's date of 1796 B.C.E. for the beginning of the Thirteenth Dynasty (as amended—see table 2.2), and dates in Bell, "Climate," 221, 225, 229, Twelfth Dynasty ruler Amenemhat III began his rule in 1858 B.C.E. According to inscriptions (Bell, 229–38), high Niles began in the second year of Amenemhat III's reign and characterized his reign. These high Niles correlate with drought conditions in Canaan: A. Issar and M. Zohar, *Climate Change*, 67, 153. The climatic situation in Canaan in the mid-nineteenth century B.C.E. would thus be in keeping with the migration of pastoralists such as Abraham from Canaan to Egypt early in Amenemhat III's reign, about 1850 B.C.E.

22. Goedicke, "Chronology," 61.

23. Griffith, "Tell el Yahudiyeh," 72. The French translation by Goyon, "Les Travaux de Chou," 31, 36, does not mention the name Thum but only the name Chou (Shu) in this particular context.

24. M. Bunson, "New Kingdom Temples," *The Encyclopedia of Ancient Egypt* (New York: Gramercy, 1991), 261.

25. Griffith, "Tell el Yahudiyeh," 72 (quote), 73.

26. Griffith, "Tell el Yahudiyeh," 72.

27. Redford, "Textual Sources," 17.

28. Cabrol, "Les Mouflons du Dieu Amon-Rê," 529–38.

29. Waddell, *Manetho*, 121–27, 137–39.

30. Goedicke, "End of the Hyksos," 43; Bietak, "Hyksos Rule," 115–17; Redford, "Coregency," 119.

31. Redford, *Egypt, Canaan, and Israel*, 414–16, and Jan Assmann, "Ancient Egyptian Anti-judaism: A Case of Distorted Memory," in *Memory Distortion*, ed. D. Schacter (Cambridge: Harvard University Press, 1997), 373, relate this story to the Amarna period late in the Eighteenth Dynasty; however there is virtually nothing that points *directly* to Amarna (a period during which Egyptians were pitted against Egyptians, not against Asiatics) and much that points to the exodus-expulsion. These exodus-expulsion elements were fused with the memory of the Hyksos and with certain bits and pieces from the early Eighteenth Dynasty to make up the story recounted by Manetho.

32. Many types of food can become contaminated with either strain of bacteria if the food comes into contact with human or animal waste during procurement or preparation.

33. D. Neev, "The Pelusium Line: A Major Transcontinental Shear," *Tectonophysics* 38 (1977), T1–T8; D. Neev, J. Hall, and J. Saul, "The Pelusium Megashear System across Africa and Associated Lineament Swarms," *JGR* 87 (1982), 1015–30.

34. For this area as a tectonic trough: D. Neev and G. M. Friedman, "Late Holocene Tectonic Activity along the Margins of the Sinai Subplate," *Science* 202 (1978): 427–28.

35. J. Hoffmeier, *Ancient Israel in Sinai: The Evidence for the Authenticity of the Wilderness Tradition* (Oxford: Oxford University Press, 2005), 94–101. He also suggests that the southwestern end of the Shi-Hor lagoon was a coastal plain through which the main outlet of the Pelusiac Nile flowed in New Kingdom times, following Bruno Marcolongo's satellite-based study of the terrain: B. Marcolongo, "Évolution du Paléo-Environment dans la Partie Orientale du Delta du Nil depuis la Transgression Flandrienne (8,000 B. P.) par Rapport aux Modèles de Peuplement Anciens," in *Sociétés Urbaines en Égypte et au Soudan*, Cahiers de Recherches de l'Institut de Papyrologie et d'Égyptologie de Lille 14 (Lille: 1992), 23–31. But Marcolongo did not ground-truth his work. Coutellier and Stanley, "Nile Delta," 257–75, based their 3500 B.P. (approximately the nineteenth century B.C.E.) coastline and Pelusiac outlet on actual sediment analysis and radiocarbon dating. Based on their northerly location of the Pelusiac outlet at 3500 B.P., I have opted for a Pelusiac outlet north of the sandstone ridge (fig. 2.1 and caption), in the eighteenth–fifteenth centuries B.C.E. Also, Marcolongo's Pelusiac branch went all the way to Pelusium, a city that began about the seventh century B.C. It's quite possible that the channel to Pelusium came into being as a result of the same tectonic activity that uplifted the Mount Casius structural ridge, shortly before 700 B.C.E.: Neev and Friedman, "Late Holocene Tectonic Activity," 427, 428 (fig. 2). Even if the southern part of the Shi-Hor were a flat coastal plain at the time Tuthmosis III crossed it in pursuit of the Israelites, the tsunamis coming through the northern Shi-Hor would cascade across this plain (see text), drowning the Egyptians.

36. See Bietak, *Avaris*, fig. 1 and Neev and Friedman, "Late Holocene Tectonic Activity," fig. 2 for this route.

37. For the fort at Tell el-Borg: J. Hoffmeier and M. Abd El-Maksoud, "A New Military Site on 'Ways of Horus'—Tell el-Borg 1999–2001: A Preliminary Report," *JEA* 89 (2003): 169–97. The channel was found by Stephen Moshier during excavation of el-Borg. J. Hoffmeier describes it as a distributary channel of the Nile (Hoffmeier, *Israel in Sinai*, 101 and n.189). The possibility also exists that this channel simply carried water from Ballah Lakes into the Shi-Hor.

38. Hoffmeier, *Israel in Sinai*, 105–07.

39. For Migdol: Hoffmeier, *Israel in Sinai*, 105–08 and figs. 5 and 10.

40. L. Hay, "What Really Happened at the Sea of Reeds?," *JBL* 83 (1964): 399.

41. Noth, *Exodus*, 102–26; F. Eakin, Jr., "The Reed Sea and Baalism," *JBL*, 379.

42. Eakin, "Baalism," 379.

43. Hay, "What Really Happened," 399.

44. Hay, "What Really Happened," 401; D. Edelman, "The Creation of Exodus 14–15," in *Jerusalem Studies in Egyptology*, ed. I. Shirun-Grumach (Wiesbaden: Harrassowitz, 1998), 139–41.

45. Edelman, "Exodus 14–15," 137 n. 2.

46. Kirsch, *Moses*, 196–97.

47. For Tuthmosis III's name: Waddell, *Manetho*, 86–87, 113. For the length of Amenophis II's reign: K. Kitchen, "The Historical Chronology of Ancient Egypt: A Current Assessment," *Acta Archaeologia* 67 (1996): 12.

48. A. Galanopoulos and E. Bacon, *Atlantis: The Truth Behind the Legend* (Indianapolis and London: Bobbs-Merrill and Nelson, 1969), 73, 193–95.

49. The 1430 B.C.E. date from traditional Greek genealogies and that from the Parian Marble is from J. Myres, *Who Were the Greeks?* (New York: Biblo and Tannen, 1967), 87, 326, 338, 340.

50. Myres, *Who Were the Greeks?*, 326.

51. Vitaliano, *Legends of the Earth*, 159.

52. For Ogyges flood and correlations of Greek flood myths with biblical events: Vitaliano, *Legends of the Earth*, 159, 259. There is also an early Deukalion tradition from northern Greece that, at least in surviving sources, is not connected to a flood but has Deukalion as the first man who makes more people out of stones. The fifth century B.C.E. Greek writer Pindar connects Deukalion with a flood, after which he and his wife come down from Parnassos (in southern Greece) and make new people out of stones. Later Greek and Roman writers have Deukalion and his wife riding out the flood in a chest: T. Gantz, *Early Greek Myth*, (Baltimore: Johns Hopkins University Press, 1993), 164–65. By the second century C.E., in the version recounted by the Roman writer Lucian, Deukalion, like the biblical Noah, rides out the flood in an ark with his wives, children, and pairs of animals. As Vitaliano (p. 158) notes: "the traditions of two different places, based on floods centuries apart, merged into what is essentially the same story."

53. Neev, Bakler, and Emery, *Mediterranean Coasts of Israel and Sinai*, 56–57, 105.

54. S. Soloviev et al., *Tsunamis in the Mediterranean Sea 2000 B.C.–2000 A.D.* (Dordrecht: Kluwer, 2000), fig. 3, lists twenty-three tsunamigenic earthquakes in the Aegean Sea.

55. For volcanic causes of tsunamis: McCoy and Heiken, "Tsunami,"1231–32.

56. For the thermoluminescence date: I. Liritzis, C. Michael, and R. Galloway, "A Significant Aegean Volcanic Eruption during the Second Millennium B.C. Revealed by Thermoluminescence Dating," *Geoarchaeology* 11 (1996): 361–71. For the submarine volcanoes between Yali and Nisyros: S. Allen, "Reconstruction of a Major Caldera-Forming Eruption from Pyroclastic Deposit Characteristics: Kos Plateau Tuff, Eastern Aegean Sea," *JVGR* 105 (2001): 143.

57. A. Sampson and I. Liritzis, "Archaeological and Archaeometrical Research at Yali, Nissiros," *TÜBA-AR* (a publication of the Turkish Academy of Sciences) 2 (1998), 104, 114 (fig. 14) describe the site on Yali and the ceramic remains, including the Minoan jug. For comparative Late Minoan IB ceramic material: Manning, *Test of Time*, 203; P. Betancourt, "Minoan Objects Excavated from Vasilike, Pseira, Sphoungaras, Priniatikos, Pyrgos, and Other Sites," *The Cretan Collection in the University Museum, University of Pennsylvania*, Vol. 1, University of Pennsylvania Museum Monograph 47 (Philadelphia, 1983), 29, 30 (numbers 56 and 57).

58. For the new radiocarbon dates on Late Minoan IA and IB and Late Minoan II: Manning et al., "Chronology for the Aegean Late Bronze Age 1700–1400 B.C.," 568–69.

59. J. Keller,"Prehistoric Pumice Tephra on Aegean Islands," *Thera III-2*, 55; Jörg Keller, letter to the author, June 11, 2002.

60. A. Sampson, "Minoan Finds from Tilos," *Athens Annals of Archaeology* 13 (1981), 68–73 (in Greek with an English abstract).

61. Jörg Keller, letter to author, June 11, 2002; Sharon Allen, e-mail to the author, June 3, 2002.

62. Noth, *Exodus*, 115 (note). The New Revised Standard Version presents Exodus 14:20 in the opposite fashion as did Martin Noth: "And so the cloud was there with the darkness, and it lit up the night."

63. M. Bunson, "Apophis," "Overthrowing Apophis," *The Encyclopedia of Ancient Egypt*, 27–28, 198; J. Taylor, *Death and the Afterlife in Ancient Egypt* (London and Chicago: University of Chicago Press, 2001), 28–29.

64. Flavius Josephus, *Antiquities of the Jews* 2, 349.

65. Taylor, *Death*, 141.

66. The Egyptian goddess Isis was the wife of the god Osiris, who was killed by his jealous brother Seth. Seth dumped Osiris's body into the Nile. Isis searched for and found it, only to have the body fall into Seth's hands. Seth cut Osiris's body into fourteen pieces and scattered it throughout Egypt. Isis searched for these parts and found all but the phallus. Once mummified, Osiris was resurrected and became the ruler of the underworld. See Taylor, *Death*, 25–28.

67. E. Wente, "Who Was Who among the Royal Mummies," *The Oriental Institute News and Notes* 144 (1995) and www-oi.uchicago.edu/OI/IS/WENTE/NN_Win95/NN_Win95.htm (9 pages, electronic version revised July 9, 2002); P. Clayton, *Chronicles of the Pharaohs*, 2nd ed. (London: Thames and Hudson, 1999), 103.

68. E. Wente, "Who Was Who among the Royal Mummies," 1, 3; W. Krogman and M. Baer, "Age at Death of Pharaohs of the New Kingdom, Determined by X-Ray Films," in *An X-Ray Atlas of the Royal Mummies*, ed. J. Harris and E. Wente (Chicago: University of Chicago Press, 1980) 202 and table 6.4; Wente, "Genealogy of the Royal Family," 122–56, and idem., "Age at Death," 238–86.

69. Wente, "Who Was Who among the Royal Mummies," 2, 8. Mitochondrial (or maternal) DNA analysis of the mummies by Scott Woodward and coworkers at Brigham Young University, Utah, has confirmed the continuation of the royal line from Sekenenre Tao to Ahmose and the break between Amenophis I and Tuthmosis I: J. Tyldesley, *The Private Lives of the Pharaohs* (New York: TV Books, 2000), 133.

70. Wente, "Who Was Who among the Royal Mummies," 4, 5–6.

71. Wente, "Who Was Who among the Royal Mummies," 4, 6; Krogman and Baer, "X-Ray Films," table 6.4; Wente, "Age at Death," 246–48.

72. Krogman and Baer, "X-Ray Films," table 6.4; Wente, "Age at Death," 251–52.

73. The discovery that a mummy at Emory University in Atlanta may be that of the Nineteenth Dynasty pharaoh Seti I leaves open the possibility that the mummy of Tuthmosis III was simply lost to tomb robbers: M. Rose, "Mystery Mummy," *Archaeology* 56, no. 2 (2003), 18–25.

74. A. Lucas, *The Route of the Exodus of the Israelites from Egypt* (London: Edward Arnold, 1938), 31–33.

Chapter 11 The Formation of the Exodus Tradition

1. For the Israelite epic tradition and spring covenant festivals: Cross, *Epic to Canon*, 29, 38–40, 43–52. The two cores of the tradition are discussed on p. 43.

2. For the Sinai peninsula forts: Oren, "Ways of Horus," 69–73. For the Egyptian presence in Canaan: idem., "'Governors' Residencies' in Canaan: A Case Study of Egyptian Administration," *Journal of the Society for the Study of Egyptian Antiquities* 14 (1984): 37–38; Mazar, *Archaeology*, 279–85; C. Higginbotham, *Egyptianization and Elite Emulation in Ramesside Palestine*, Vol. 2 of *Culture and History of the Ancient Near East* (Leiden: Brill, 2000), 129–38. Egyptian administrative centers appear at Gaza, Jaffa, and Beth Shean: I. Singer, "Merneptah's Campaign to Canaan and the Egyptian Occupation of the Southern Coastal Plain of Palestine in the Ramesside Period," *BASOR* 269 (1988): 1–9.

3. For Papyrus Anastasi V: R. Caminos, *Late-Egyptian Miscellanies* (London: G. Cumberledge, 1954), 255.

4. Finkelstein, *Settlement*, 309–11.

5. Finkelstein, *Settlement*, 324–29; Stager, "Song of Deborah," 55–57.

6. Gal, "Lower Galilee," 35–46; Finkelstein, *Settlement*, 94–97; Frankel, "Upper Galilee," 26–29, 33–34.

7. Ofer, "Hill Country of Judah," 108–09, 118–19.

8. W. Dever, "Save Us from Postmodern Malarkey," *BAR* 26 (March–April 2000), 32.

9. Pennington, "Early Human Population Growth," 259–74 (quote 259); see especially table 3 and its accompanying discussion, which demonstrates that populations starting with only 1,000 individuals increased to 81,000–181,000 in two hundred years, despite increased adult mortality rates.

10. Magny, "Oceanic and Solar Forcing," 280 (fig. 1), 282 (fig. 3); J. Neumann and S. Parpola, "Climatic Change and the Eleventh-Tenth-Century Eclipse of Assyria and Babylonia," *JNES* 46 (1987): 162, 169–71.

11. Maley, "Middle to Late Holocene Changes," 617–19; K. Butzer, "Long-Term Nile Flood Variation and Political Discontinuities in Pharaonic Egypt," In *From Hunters to Farmers: The Causes and Consequences of Food Production in Africa*, ed. J. Clark and S. Brandt (Berkeley: University of California Press, 1984), 107.

12. Frumkin, "Holocene History of Dead Sea Levels," 243.

13. Neumann and Parpola, "Climatic Changes," 163–64.

14. B. Weiss, "The Decline of Late Bronze Age Civilization as a possible response to Climatic Change," *Climatic Change* 4 (1982): 183.

15. Neumann and Parpola, "Climatic Changes," 161, 171–81; Na'aman, "Conquest of Canaan," 243–44; Hassan, "Nile Floods," 7; Stanley et al. "Nile Flow Failure," 399.

16. F. Yurco, "Merenptah's Canaanite Campaign," *Journal of the American Research Center in Egypt* 23 (1986): 189–215.

17. Singer, "Egyptians, Canaanites, and Philistines," 290–93; Weinstein, "Egyptian Empire," 18–21.

18. Na'aman, "Conquest of Canaan," 243–44; Bunimovitz, "Central Hill Country," 201 and n. 110.

19. The Israelites themselves, of course, may have had something to do with the destruction of some of these cities: Singer, "Egyptians, Canaanites, and Philistines," 284, 291.

20. Finkelstein, *Settlement*, 238–59, 264–67; L.Stager, "The Archaeology of the Family in Ancient Israel," *BASOR* 260 (1985), 11–17.

21. In the central highlands, an early form of cooking pot and a small jug are clearly derived from an earlier Late Bronze Age form while in Upper Galilee the "Galilean" type of pithoi jar is derived from a Late Bronze Age type found at Canaanite Hazor: Zertel, "Land of the Perizzites," 52–54; Frankel, "Upper Galilee," 27–29. See A. Zertel, "Philistine Kin Found in Israel," *BAR* 28 (May–June 2002), 23–24, for a brief summary of the differences between highland and Canaanite pottery. As a whole, Iron I pottery shows more regional diversity than did preceding Late Bronze Age styles and is typified by the "collared-rim" storage jar from the Central Highlands, Lower Galilee, and at Dan in the far north: Finkelstein, *Settlement*, 102–03, 108–09, 313, 322, 337.

22. As Baruch Halpern pointed out (*The Emergence of Israel in Canaan*. Society of Biblical Literature Monograph Series 29 (Chico, Calif.: Scholars Press, 1983), 99, 103), at least as early as the twelfth century B.C.E., and particularly noted in the Song of Deborah (Judges 5:19), Israel distinguished itself from the surrounding Canaanite peoples on the basis of ethnicity.

23. A. Nur and E. Cline, "Poseidon's Horses: Plate Tectonics and Earthquake Storms in the Late Bronze Age Aegean and Eastern Mediterranean," *Journal of Archaeological Science* 27 (2000): 43–63.

24. Mazar, *Archaeology*, 307–08.

25. L. Stager, "The Impact of the Sea Peoples in Canaan (1185–1050 BCE)," *ASHL*, 340–44.

26. Egyptian grain prices are quoted in: Butzer, "Long-Term Nile Flood Variation," 109–10.

27. Stager, "Sea Peoples," 336 (fig. 2), 343; Weinstein, "Egyptian Empire," 22–23.

28. For an excellent discussion of the ecological, socioeconomic, and demographic aspects of Israelite growth and expansion westward from the central hills and desert fringe out onto the slopes and lowlands in the eleventh century B.C.E. see I. Finkelstein, "The Emergence of the Monarchy in Israel: The Environmental and Socio-Economic Aspects," *Journal for the Study of the Old Testament* 44 (1989): 43–74.

29. For the destruction of Shiloh after the Battle of Ebenezer: Finkelstein, *Settlement*, 205, 220–28. For Josephus and his numbers: Chapter 4.

30. A. Malamat, "The Proto-History of Israel: A Study in Method," in *The Word of the Lord Shall Go Forth: Essays in Honor of David Noel Freedman*, ed. C. Meyers and M. O'Connor (Winona Lake, Ind.: Eisenbrauns, 1983), 303–05.

Appendix Oral Transmission, Memory and Recall, and Oral History

1. J. Vansina, *Tradition*, 4–6.

2. F. Bartlett, *Remembering: A Study in Experimental and Social Psychology* (Cambridge: Cambridge University Press, 1932), 93.

3. Bartlett, *Remembering*, 175.

4. G. Stephenson, H. Brandstätter, and W. Wagner, "An Experimental Study of Social Performance and Delay on the Testimonial Validity of Story Recall," *European Journal of Social Psychology* 13 (1983): 175–91 (quote, 188); D. Edwards and D. Middleton,

"Joint Remembering: Constructing an Account of Shared Experience through Conversational Discourse," *Discourse Processes* 9 (1986): 423–59; W. Hirst and D. Manier, "Remembering as Communication: A Family Recounts Its Past," in *Remembering Our Past: Studies in Autobiographical Memory*, ed. D. Rubin (Cambridge: Cambridge University Press, 1996), 271–90.

5. Stephenson, Brandstätter, and Wagner, "Experimental Study," 177–79, 187.

6. Brewer, "Autobiographical Events," 38, 48; M. Linton, "Transformations of Memory in Everyday Life," in *Memory Observed: Remembering in Natural Contexts*, ed. U. Neisser (San Francisco: Freeman, 1982), 86, 90; R. White, "Memory for Personal Events," *Human Learning* 1 (1982): 17.

7. E. Loftus and W. Marburger, "Since the Eruption of Mt. St. Helens, Has Anyone Beaten You Up? Improving the Accuracy of Retrospective Reports with Landmark Events," *Memory and Cognition* 11 (1983), 114–120.

8. For one study of this phenomenon: C. Thompson, J. Skowronski, and D. Lee, "Telescoping in Dating Naturally Occurring Events," *Memory and Cognition* 16 (1988): 461–68.

9. Allport and Postman, *Rumor*, 65–73.

10. R. Avigdor, "The Psychology of Rumor," *Psychological Newsletter* 3 (1951): 3; F. Brissey, "The Factor of Relevance in the Serial Reproduction of Information," *Journal of Communication* 11 (1961): 218.

11. D. Campbell, "Systematic Error on the Part of Human Links in Communication Systems," *Information and Control* 1 (1958), 343.

12. Allport and Postman, *Rumor*, 86–98.

13. Allport and Postman, *Rumor*, 101–05; Higham, "Rumour," 42–55; Campbell, "Systematic Error," 347–52.

14. Higham, "Rumour," 51.

15. J. Vansina, "Memory and Oral Tradition," in *The African Past Speaks: Essays on Oral Tradition and History*, ed. J. C. Miller (Hamden, Conn.: Archon, 1980), 263.

16. Vansina, *Tradition*, 171.

17. Hirst and Manier, "Remembering," 281; R. Wilson, *Genealogy and History in the Biblical World* (New Haven: Yale University Press, 1977), 29–30; Vansina, *Tradition*, 100.

18. Vansina, *Tradition*, 23–24, 168–69; D. Henige, *Oral Historiography* (London: Longman, 1982), 100.

Glossary of Geological and Technical Terms

(These terms are found underlined in their first appearance in the text.)

Acid spikes: Highly acidic layers found within an ice core. See also ice cores.

Alkaline basalts: Volcanic rocks that are rich in the minerals plagioclase feldspar and pyroxene, and that occur primarily in lava flows. They are high in the elements sodium (Na) and potassium (K) and lower in silica (SiO_2) and calcium (Ca) in comparison to other types of volcanic rocks, and are basic rather than acidic.

Animal vector: Any animal that acts as a carrier or agent of a disease-causing organism such as a virus or a bacillus.

Asthenosphere: The layer within the earth's mantle directly below the lithosphere. It begins about 100 km below the earth's surface. In this region temperatures and pressures are high enough to cause rock to be partially melted.

Caldera: A large, circular depression or crater formed by the collapse of the top of a volcanic cone, usually because the underlying magma chamber has been rapidly emptied by an explosive eruption.

Calibration curve: A graph with the radiocarbon age (see below) plotted against the calendar age, based on the variations of atmospheric ^{14}C determined from tree rings (see dendrochronology). The curve contains wiggles that represent short-term variations in the amount of ^{14}C in the atmosphere through time, so that many radiocarbon dates have more than one calendar age. Computer programs are often used instead of "wiggle-matching" the graph itself. These programs give statistical probabilities so that the most likely calendar date range can be determined for a radiocarbon date.

Co-ignimbrite eruption column: The buoyant column or plume that forms when pyroclastic flows (see below) from an erupting volcano heat the air within and above themselves, causing the air to rise and carry volcanic dust, ash, and volcanic gases with it, so massive amounts of erupted material penetrate high into the atmosphere. Material from a co-ignimbrite eruption column will not reach as high into the atmosphere as material from a plinian eruption (see below).

Dendrochronology: See Tree-ring chronology.

Diatoms: Certain types of one-celled algae that have a convoluted shell made of silica (SiO_2).

Effusive eruption: A volcanic eruption in which lava flows along the surface of the earth.

Explosive eruption: A volcanic eruption which ejects gases, ash and other volcanic material.

Fissure: A linear volcanic vent or opening through which gases or lava escape.

Ice cores: Each year the snow falling on the polar ice packs leaves its own distinct layer, which can be detected even after the layers become compacted into ice. This ice contains a climatic record going back 200,000 years, a record that can be recovered by drilling, removal, and study of an ice core. These cores have been brought back to laboratories in the United States and Europe for study. The three principal ice cores drilled into the Greenland ice cap are known as the GRIP, GISP2, and Dye-3 ice cores. Other cores have been drilled into Antarctic ice. In up to three years after acid aerosols from a volcanic eruption reach the stratosphere, they will fall back down to earth and be deposited with snow on the ice packs. Even minute amounts of acid can be detected in these layers in the ice core, because the acid increases the conductivity of the ice, and electrodes can be used to detect this conductivity. A highly acidic layer is sometimes called an acid spike. After an acid layer is detected, it can be analyzed chemically to determine if it is likely to have come from a volcanic source.

Ignimbrites: Rocks formed by the consolidation of hot volcanic material that makes up ash flows or pyroclastic flows (see below).

Intertropical Convergence Zone (ITCZ): The dividing line between the northwest and the southeast trade winds, sometimes called the meteorological equator; this line or band shifts according to the seasons and has a decisive effect on rainfall patterns in the tropics, particularly that of monsoonal rainfall.

Lineament: A linear topographic feature that extends for a great distance and reflects some underlying structure in the earth.

Magma: Molten rock that originates in the earth's mantle (see below) and crust.

Magma chamber: A large chamber or reservoir within the earth's crust containing magma that has risen from much deeper in the earth's mantle.

Mantle: The section of the earth's interior between the outer crust and the inner core.

Mantle-crust boundary: The seismic discontinuity (sometimes known as the Mohorovičić discontinuity or Moho) that lies about thirty-five kilometers below the continents and ten kilomenters beneath the oceans.

Normal fault: A fracture or fracture zone between two blocks, one of which, known as the hanging wall, appears to have slipped downward relative to the other side or block, known as the footwall. This type of fault is found throughout the world.

Olivine phenocrysts: Olivine is a common rock-forming mineral, $(Mg,Fe)_2SiO_4$, that is often found as larger crystals within finer-grained igneous rocks such as basalts. These crystals were formed as the rock cooled from the molten magma.

Paleosol: An ancient soil, often buried by more recent soils or sediments.

Plinian eruption: Sometimes called a vulcanian eruption, this type of volcanic eruption is characterized by periodic explosive events in which eruptive material shoots upward in a high towering column of gas and ash.

Pull-apart basin: A special type of basin associated with an intracontinental transform valley (see tectonic rift). En echelon strike-slip faults moving in opposite directions parallel to the long axis of the valley produce a long rectangular basin that grows lengthwise as faulting continues.

Pumice: A highly porous volcanic rock formed from cooling lava that is filled with expanding gas bubbles.

Pyroclastic flows: Sometimes called *nuées ardentes*, these clouds of hot, glowing (incandescent) volcanic gases and suspended ash are erupted from a volcano and can travel at great speeds along the surface of either the ground or the sea. Sometimes a distinction is made between surges (ash clouds) and flows (heavier, semiliquid material).

Pyroclastic: Literally, "fire-broken," this term refers to any type of rock formed from explosive volcanic eruptions.

Radiocarbon dating: Radiocarbon dating measures the decay of ^{14}C, the radioactive isotope of carbon produced by cosmic rays in the atmosphere. Just like the two other isotopes of carbon (^{12}C and ^{13}C), ^{14}C combines with oxygen in the atmosphere to form carbon dioxide gas (CO_2), and in this molecular form enters all living organisms. While an organism, plant or animal, is alive, the amount of ^{14}C it contains reflects the amount of radioactive carbon in the atmosphere at that time. The variation in abundance of ^{14}C in the atmosphere through time can be determined by measuring the yearly accumulations of ^{14}C that are found in tree rings, which themselves, when sequenced, can give calendar dates back through time. Once an organism dies or stops growing, carbon no longer passes through its body and the ^{14}C stays in place, decaying at a constant rate over time. Because this decay rate (its "half-life") is known, as is the normal amount of ^{14}C in the atmosphere at any particular time, comparing the amount of ^{14}C relative to the two other isotopes of carbon in a sample of organic material can determine the date of the organism's death (or when it stopped growing) and thus date the sample. The latest advance in radiocarbon dating is the use of accelerator mass spectrometry (AMS) to measure ^{14}C.

Radiocarbon Years Before Present: The number of years before 1950 C.E. calculated by the half-life of ^{14}C (which is 5,780 years). These years must be converted by a calibration program or calibration curve (see above) to get a calendar year.

Secondary maxima: The depth and mass of a volcanic ashfall typically lessens the farther the ash has traveled through the air before it falls to the ground. Sometimes, however, areas of anomalously substantial ashfall are found more than 150 kilometers from an eruption. These areas are called secondary maxima or secondary mass maxima.

Seismic waves: Shock waves generated by earthquakes.

Structural ridge: A ridge produced by the displacement or deformation of rocks.

Subduction: The downward movement or emplacement of a tectonic plate into the mantle beneath an overriding tectonic plate.

Tectonic plates: The rigid blocks of the earth's crust that form the rocky surface of the earth. These blocks move against each other in various ways as new crust is formed along the mid-ocean ridges and older crust is consumed in subduction zones.

Tectonic rift: Rifts are thought to represent areas where a tectonic plate is separating or breaking up. Rifts on land usually occur on areas of uplifted crust when hotter magma stretches the earth's crust and the central part of the crust or dome collapses, forming a steep-sided valley or transform valley delimited by parallel faults on each side of the low-lying floor.

Tephra: Any type of volcanic material that is explosively ejected during a volcanic eruption.

Thermoluminescence dating: Electrons and light are released from the crystal structures of certain materials such as pottery, clay, or quartz sands when these materials have been heated or exposed to sunlight. This released light can be measured and used to estimate the time elapsed since the object was last heated. Unfortunately, this is usually not a very precise form of dating.

Tree-ring chronology: Sometimes called dendrochronology or tree-ring dating, the study of tree rings in order to date past events. Annual growth rings can be counted to date wood, and the variant widths of these rings can be measured and correlated with those from other trees as well as matched against established sequences that will produce an exact calendar year date. Annual rings whose calendar date has been determined can be measured for their ^{14}C content. This will provide a record of the variation in atmospheric ^{14}C through time and give an accurate calibration for radiocarbon dating.

Tsunami: A particularly large sea wave produced when an earthquake, undersea landslide, or volcanic eruption displaces a large volume of seawater, thus causing the wave to form.

Turbidity: Not clear, muddied, or disturbed, as in the case of water that contains sediment which has been stirred up.

Volcanic Explosivity Index (V.E.I.): The scale used to classify or measure volcanic eruptions, it is based on the amount of material ejected in an eruption and the height the eruptive material reaches in the atmosphere. The highest known V.E.I. is an 8, used for the Toba eruption of 74,000 years ago.

Bibliography

Abd El-Maksoud, M. *Tell Heboua (1981–1991): Enquête archéologique sur la Deuxième Période Intermédiaire et le Nouvel Empire à l'exteémité orientale du Delta.* Paris: Éditions Recherche sur les Civilisations, 1998.

Aberbach, M., and L. Smolar. "Aaron, Jeroboam and the Golden Calves." *JBL* 86 (1967): 129–40.

Albright, W. "Northwest-Semitic Names in a List of Egyptian Slaves from the Eighteenth Century B.C." *Journal of the American Oriental Society* 74 (1954): 222–32.

Albright, W. "Midianite Donkey Caravans." In *Translating and Understanding the Old Testament,* edited by H. Frank and W. Reed, 197–205. Nashville: Abingdon, 1970.

Aharoni, Y. *The Land of the Bible: A Historical Geography,* translated by A. Rainey. Philadelphia: Westminster, 1979.

Allen, S. "Reconstruction of a Major Caldera-Forming Eruption from Pyroclastic Deposit Characteristics: Kos Plateau Tuff, Eastern Aegean Sea." *JVGR* 105 (2001): 141–62.

Allport, G., and L. Postman. *The Psychology of Rumor.* New York: Holt, 1947.

Almond, D. "Geological Evolution of the Afro-Arabian Dome." *Tectonophysics* 131 (1986): 301–32.

Almond, D. "The Relation of Mesozoic-Cainozoic Volcanism to Tectonics in the Afro-Arabian Dome." *JVGR* 28 (1986): 225–46.

Anbar, M., and N. Na'aman. "An Account Tablet of Sheep from Ancient Hebron." *Tel Aviv* 13–14 (1986–1987): 3–12.

Appleby, A. "The Disappearance of Plague: A Continuing Puzzle." *Economic History Review* 33 (1980): 161–73.

Assmann, J. "Ancient Egyptian Antijudaism: A Case of Distorted Memory." In *Memory Distortion,* edited by D. Schacter, 365–76. Cambridge: Harvard University Press, 1992.

Avigdor, R. "The Psychology of Rumor." *Psychological Newsletter* 3 (1951): 6–10.

Avishur, Y. "The Narrative of the Revelation at Sinai (Ex 19–24)." In *Studies in Historical Geography and Biblical Historiography presented to Zecharia Kallai,* 197–214. Supplements to Vetus Testamentum, no. 81. Leiden: Brill, 2000.

Avner, U. "Sacred Stones in the Desert." *BAR* 27 (May–June 2001): 30–41.

Awadallah, A., and J. Rousselle. "Improving Forecasts of Nile Flood Using SST Inputs in TFN Model." *Journal of Hydrologic Engineering* 5 (2000): 371–79.

Barber, E. W., and P. T. Barber. *When They Severed Earth from Sky: How the Human Mind Shapes Myth.* Princeton: Princeton University Press, 2004.

Bar-Deroma, H. "Kadesh-Barne'a." *PEQ* 96 (1964): 101–34.

Bar-Matthews, M., A. Ayalon, A. Kaufman, and G. Wasserburg. "The Eastern Mediterranean Paleoclimate as a Reflection of Regional Events: Soreq Cave, Israel." *Earth and Planetary Science Letters* 166 (1999): 85–95.

Bartlett, F. *Remembering: A Study in Experimental and Social Psychology* (Cambridge: Cambridge University Press, 1932.

Bauer, P. "Review of Hekla: A Notorious Volcano by Sigurdur Thorarinsson," *Science* 172 (1971): 692–93.

Baxter, P., R. Ing, H. Falk, and B. Plikaytis. "Mount St. Helens Eruptions: The Acute Respiratory Effects of Volcanic Ash in a North American Community," *Archives of Environmental Health* 38, no. 3 (1983): 138–43.

Bayer, H.-J., H. Hötzl, A. Jado, B. Roscher, and W. Voggenreiter. "Sedimentary and Structural Evolution of the Northwest Arabian Red Sea Margin," *Tectonophysics* 153 (1988): 137–51.

Bell, B. "Climate and the History of Egypt: The Middle Kingdom." *American Journal of Archaeology* 79 (1975): 223–69.

Ben-Menahem, A. "Four Thousand Years of Seismicity along the Dead Sea Rift." *JGR* 96 (1991): 20195–216.

Bennett, C.-M. "Biblical Traditions and Archaeological Results." In *The Archaeology of Jordan and Other Studies Presented to Siegfried H. Horn*, edited by L. Geraty and L. Herr, 75–83. Berrien Springs: Andrews University Press, 1986.

Bennett, J. "Geo-physics and Human History: New Light on Plato's Atlantis and the Exodus." *Systematics* 1 (1963): 127–56.

Benson, D. *Ancient Egypt's Warfare: A Survey of Armed Conflict in the Chronology of Ancient Egypt 1600 B.C.–30 B.C.* Chicago: D. Benson, 1995.

Ben-Tor, D., S. Allen, and J. Allen. "Seals and Kings." *BASOR* 315 (1999): 47–74.

Betancourt, P. "Minoan Objects Excavated from Vasilike, Pseira, Sphoungaras, Priniatikos, Pyrgos, and Other Sites." *The Cretan Collection in the University Museum, University of Pennsylvania.* Vol. 1. University of Pennsylvania Museum Monograph, no. 47. Philadelphia, 1983.

Betancourt, P., and G. Weinstein. "Carbon-14 and the beginning of the Late Bronze Age in the Aegean." *American Journal of Archaeology* 80 (1976): 329–48.

Bienkowski, P. *Jericho in the Late Bronze Age.* Warminster: Aris and Phillips, 1986.

Bietak, M. *Tell el-Dab'a II. Der Fundort im Rahmen, einer archäologisch-geographischen Untersuchung über das ägyptische Ostdelta.* Denkschriften der Gesamtakademie, no. 4. Vienna: Österreichische Akademie der Wissenschaften, 1975.

Bietak, M. *Avaris and Piramesse: Archaeological Exploration in the Eastern Nile Delta.* Proceedings of the British Academy, no. 65. Oxford: Oxford University Press, 1979.

Bietak, M. *Avaris: The Capital of the Hyksos.* London: British Museum Press, 1996.

Bietak, M. "The Center of Hyksos Rule: Avaris (Tell el-Dab'a)." *Hyksos*, 87–139.

Bietak, M. "The Volcano Explains Everything—Or Does It?" *BAR* 32 (November–December 2006): 60–65.

Biggs, R. "Medicine, Surgery, and Public Health in Ancient Mesopotamia." In *Civilizations of the Ancient Near East*, Vol. 3, edited by J. Sasson, 1911–24. New York: Scribners', 1995.

Bimson, J. *Redating the Exodus and Conquest*. Sheffield: Almond, 1981.

Bimson, J., and D. Livingston. "Redating the Exodus," *BAR* 13 (September–October 1987): 40–53, 66–68.

Biraben, J.-N., *Les Hommes et la Peste en France et dans les Pays Européens et Mediterranéens*, Vol. 1. Paris: Mouton, 1975.

Bircher, W. "Dendrochronology Applied in Mountain Regions." In *Handbook of Holocene Palaeoecology and Palaeohydrology*, edited by B. Berglund, 387–403. Chichester: Wiley-Interscience, 1986.

Bitschene, P., and H.-U. Schmincke. "Fallout Tephra Layers: Composition and Significance." In *Sediments and Environmental Geochemistry: Selected Aspects and Case Histories*, edited by D. Heling, P. Rothe, U. Förstner, and P. Stoffers, 48–82. Berlin: Springer, 1990.

Blong, R. "Time of Darkness Legends from Papua New Guinea." *Oral History* 7 (1979): 1–135.

Blong, R. *The Time of Darkness: Local Legends and Volcanic Reality in Papua New Guinea*. Seattle: University of Washington Press, 1982.

Blong, R. *Volcanic Hazards*. Sydney: Academic Press, 1984.

Bolt, B. *Earthquakes*. New York: Freeman, 1988.

Bond, A., and R. Sparks. "The Minoan Eruption of Santorini, Greece." *Journal of the Geological Society of London* 132 (1976): 1–16.

Bookman, R., Y. Enzel, A. Agnon, and M. Stein. "Late Holocene Lake Levels of the Dead Sea." *GSABull* 116 (2005): 555–71.

Bourriau, J. "Relations between Egypt and Kerma during the Middle and New Kingdoms." In *Egypt and Africa: Nubia from Prehistory to Islam*, edited by W. Davies, 129–44. London: British Museum, 1991.

Bowman, S., J. Ambers, and M. Leese. "Re-evaluation of British Museum Radiocarbon Dates Issued between 1980 and 1984." *Radiocarbon* 32 (1990): 59–79.

Breasted, J. *Ancient Records of Egypt: Historical Documents*. Vol. 2, *The Eighteenth Dynasty*. Chicago: University of Chicago Press, 1906.

Brewer, W. "Memory for Randomly Sampled Autobiographical Events." In *Remembering Reconsidered*, edited by U. Neisser and E. Winograd, 21–90. Cambridge: Cambridge University Press, 1988.

Brewer, W. "The Theoretical and Empirical Status of the Flashbulb Memory Hypothesis." In *Affect and Accuracy in Recall: Studies of "Flashbulb" Memories*, edited by E. Winograd and U. Neisser, 274–305. Cambridge: Cambridge University Press, 1992.

Briffa, K., P. Jones, F. Schweingruber, and T. Osborn. "Influence of Volcanic Eruptions on Northern Hemisphere Summer Temperature over the Past 600 Years." *Nature* 393 (1998): 450–55.

Brissey, F. "The Factor of Relevance in the Serial Reproduction of Information." *Journal of Communication* 11 (1961): 211–19.

Brown, G., R. Jackson, R. Bogue, and E. Elberg, Jr., "Geologic Map of the Northwestern Hijaz Quadrangle, Kingdom of Saudi Arabia." *U.S. Geological Survey Miscellaneous Geologic Investigations Map I-204 A*. Scale 1:500,000. Washington: U.S. Geological Survey, 1963.

Brown, G., D. Schmidt, and A. Huffman, Jr., *Geology of the Arabian Peninsula: Shield Area of Western Saudi Arabia*. U.S. Geological Survey Professional Paper 560-A. Washington: U.S. Government Printing Office, 1989.

Bruce-Chwatt, L. "Malaria in African Infants and Children in Southern Nigeria." *Annals of Tropical Medicine and Parasitology* 46 (1952): 173–200.

Bruins, H., and J. van der Plicht. "The Exodus Enigma." *Nature* 382 (1996): 213–14.

Bruner, J., and C. Feldman. "Group Narrative as a Cultural Context of Autobiography." In *Remembering our Past: Studies in Autobiographical Memory*, edited by D. Rubin, 291–317. Cambridge: Cambridge University Press, 1996.

Bryce, T., 1998, *The Kingdom of the Hittites*. Oxford: Clarendon Press, 1998.

Bunimovitz, S. "Socio-Political Transformations in the Central Hill Country in the Late Bronze–Iron I Transition." *FNM*, 179–202.

Bunimovitz, S. "On the Edge of Empires—Late Bronze Age (1500–1200 BCE)." *ASHL*, 320–31.

Bunson, M. *The Encyclopedia of Ancient Egypt*. New York: Gramercy Books, 1991.

Burton, I., R. Kates, and G. White. *The Environment as Hazard*. 2nd ed. New York: Guildford, 1993.

Butler, T. "Yersiniosis and Plague." In *Zoonoses: Biology, Clinical Practice and Public Health Control*, edited by S. Palmer, L. Soulsby, and D. Simpson, 281–93. Oxford: Oxford University Press, 1998.

Butzer, K. *Early Hydraulic Civilizations in Egypt*. Chicago: University of Chicago Press, 1976.

Butzer, K. "The Holocene Lake Plain of North Rudolph, East Africa." *Physical Geography* 1 (1980): 42–58.

Butzer, K. "Long-Term Nile Flood Variation and Political Discontinuities in Pharaonic Egypt." In *From Hunters to Farmers: The Causes and Consequences of Food Production in Africa*, edited by J. Clark and S. Brandt, 102–12. Berkeley: University of California Press, 1984.

Cabrol, A. "Les Mouflons du Dieu Amon-Rê." In *Egyptian Religion: The Last Thousand Years*, edited by W. Clarysse, A. Schoors, and H. Willems, 529–38. Leuven: Uitgeverij Peeters and Dept. Oosterse Studies, 1998.

Callaway, J. "Excavating at Ai (Et-Tell): 1964–1972." *Biblical Archaeologist* 39 (1976): 18–30.

Caminos, R. *Late-Egyptian Miscellanies*. London: G. Cumberledge, 1954.

Camp, V., and M. Roobol. "The Arabian Continental Alkali Basalt Province: Part I. Evolution of Harrat Rahat, Kingdom of Saudi Arabia." *GSABull* 101 (1989): 71–95.

Camp, V., and M. Roobol. "Upwelling Asthenosphere beneath Western Arabia and Its Regional Implications." *JGR* 97 (1992): 15,255–71.

Campbell, D. "Systematic Error on the Part of Human Links in Communication Systems." *Information and Control* 1 (1958): 334–69.

Casperson, L. "The Lunar Dates of Thutmose III." *JNES* 45 (1986): 139–50.

Cassan, L. *The Ancient Mariners*. 2nd ed. Princeton: Princeton University Press, 1991.

Cavanaugh, D. "Specific Effect of Temperature upon Transmission of the Plague Bacillus by the Oriental Rat Flea, *Xenopsylla cheopis*." *American Journal of Tropical Medicine and Hygiene* 20 (1971): 264–73.

Chadwick, J. "Discovering Hebron: The City of the Patriarchs Slowly Yields Its Secrets." *BAR* 31 (September–October 2005): 24–33, 70–71.

Childs, B. "The Birth of Moses."*JBL* 84 (1965): 109–22.

Childs, B. *The Book of Exodus: A Critical, Theological Commentary.* Philadelphia: Westminster, 1974.

Christiansen, R. "Epic Movie Adventures with Charlton Heston," *Chicago Tribune,* March 21, 2001.

Christie, A., T. Chen, and S. Elberg. "Plague in Camels and Goats: Their Role in Human Epidemics." *Journal of Infectious Diseases* 141 (1980): 724–26.

Cita, M., A. Camerlenghi, K. Kastens, and F. McCoy, "New Findings of Bronze Age Homogenites in the Ionian Sea: Geodynamic Implications for the Mediterranean." *Marine Geology* 55 (1984): 47–62

Clark, E. *Indian Legends of the Pacific Northwest.* Berkeley: University of California Press, 1953.

Clastres, H. *The Land-Without-Evil: Tupí-Guaraní Prophetism.* Urbana: University of Illinois Press, 1995.

Clausen, H., C. Hammer, C. Hvidberg, D. Dahl-Jensen, J. Steffensen, J. Kipfstuhl, and M. Legrand. "A Comparison of the Volcanic Records over the Past 4000 Years from the Greenland Ice Core Project and Dye 3 Greenland Ice Cores." *JGR* 102 (1997): 26707–23.

Clayton, P. *Chronicle of the Pharaohs.* 2nd ed. London: Thames and Hudson, 1999.

Coates, G. "Despoiling the Egyptians." *Vetus Testamentum* 18 (1968): 450–57.

Coates, G. *Rebellion in the Wilderness.* Nashville: Abingdon, 1968.

Coates, G. *Moses: Heroic Man, Man of God.* Sheffield: JSOT Press, 1988.

Cohen, R. "Did I Excavate Kadesh-Barnea?" *BAR* 7 (May–June 1981): 20–33.

Cole-Dai, J., E. Mosley-Thompson, and L. Thompson. "Annually Resolved Southern Hemisphere Volcanic History from Two Antarctic Ice Cores." *JGR* 102 (1997): 16,761–71.

Coleman, S. *Volcanoes New and Old.* New York: John Day, 1946.

Collins, J. "Artapanus (Third to Second Century b.c.): A New Translation and Introduction." In *The Old Testament Pseudepigrapha,* Vol. 2, edited by J. Charlesworth, 889–903. Garden City: Doubleday, 1985.

Cook, R., J. Barron, R. Papendick, and G. Williams. "Impact on Agriculture of the Mount St. Helens Eruptions." *Science* 211 (1981): 16–22.

Coutellier, V., and D.-J. Stanley. "Late Quaternary Stratigraphy and Paleogeography of the Eastern Nile Delta, Egypt." *Marine Geology* 77 (1987): 257–75.

Cross, F., Jr. *Canaanite Myth and Hebrew Epic.* Cambridge: Harvard University Press, 1973.

Cross, F., Jr. *From Epic to Canon: History and Literature in Ancient Israel.* Baltimore: Johns Hopkins University Press, 1998.

Cullen, H., and P. de Menocal. "North Atlantic Influence on Tigris-Euphrates Streamflow." *International Journal of Climatology* 20 (2000): 853–63.

D'Arrigo, R., D. Frank, G. Jacoby, and N. Pederson. "Spatial Response to Major Volcanic Events in or About AD 536, 934 and 1258: Frost Rings and Other Dendrochronological Evidence from Mongolia and Northern Siberia: Comment on R. Strothers,

'Volcanic Dry Fogs, Climate Cooling, and Plague Pandemics in Europe and the Middle East.'" *Climatic Change* 49 (2001): 239–46.

Davies, G. *The Way of the Wilderness*. Cambridge: Cambridge University Press, 1979.

Davies, G. "The Wilderness Itineraries and Recent Archaeological Research." In *Studies in the Pentateuch*, edited by J. Emerton, 161–75. Vetus Testamentum Supplement, no. 41. Leiden: Brill, 1990.

DeMallie, R., and D. Parks. "Tribal Traditions and Records." In *Handbook of North American Indians*, Vol. 13, edited by R. DeMallie, 1062–73. Washington: Smithsonian, 2001.

de Silva, S., and J. Alzueta. "The Socioeconomic Consequences of the A.D. 1600 Eruption of Huaynaputina, Southern Peru." In *Volcanic Hazards and Disasters in Human Antiquity*, edited by F. McCoy and G. Heiken, 15–24. Geological Society of America Special Paper, no. 345. Boulder, Co., 2000.

de Silva, S., and G. Zielinski. "Global Influence of the AD 1600 Eruption of Huaynaputina, Peru." *Nature* 393 (1998): 455–58.

de Vaux, R. *The Early History of Israel*. Philadelphia: Westminster, 1978.

Dever, W. "Palestine in the Second Millennium BCE: The Archaeological Picture." *IJH* 70–120.

Dever, W. "Relations between Syria-Palestine and Egypt in the 'Hyksos' Period," In *Palestine in the Bronze and Iron Ages: Papers in Honor of Olga Tufnell*, edited by J. Tubb, 69–87. London: Institute of Archaeology Press, 1985.

Dever, W. "The Chronology of Syria-Palestine in the Second Millennium B.C.E.: A Review of Current Issues." *BASOR* 288 (1992): 1–25.

Dever, W. G. "Is There Any Archaeological Evidence for the Exodus?" In *Exodus: The Egyptian Evidence*, edited by E. S. Frerichs and L. H. Lesko, 67–86. Winona Lake, Ind.: Eisenbrauns, 1997.

Dever, W. G. "Settlement Patterns and Chronology of Palestine in the Middle Bronze Age." *Hyksos*, 285–301.

Dever, W. "Nelson Glueck and the Other Half of the Holy Land." In *The Archaeology of Jordan and Beyond*, edited by L. Stager, J. Greene, and M. Coogan, 114–21. Winona Lake, Ind.: Eisenbrauns, 2000.

Dever, W. "Save Us from Postmodern Malarkey." *BAR* 26 (March–April 2000), 28–35, 68–69.

Doughty, C. *Travels in Arabia Deserta*, Vol. 1. London: Jonathan Cape, 1923.

Doumas, C. *Thera: Pompeii of the Ancient Aegean*. London: Thames and Hudson, 1983.

Eakin, F., Jr., "The Reed Sea and Baalism." *JBL* 86 (1967): 378–84.

Eastwood, W., J. Tibby, N. Roberts, H. Birks, and H. Lamb. "The Environmental Impact of the Minoan Eruption of Santorini (Thera): Statistical Analysis of Palaeo-ecological Data from Gölhisar, Southwest Turkey." *The Holocene* 12 (2002): 431–44.

Ebbinghaus, H. *Memory: A Contribution to Experimental Psychology*. New York: Columbia University Teachers College Press, 1913.

Eddy, J., "Historical and Arboreal Evidence for a Changing Sun." In *The New Solar Physics*, edited by J. Eddy, 11–33. Boulder, Co.: Westview Press, 1978.

Edelman, D. "The Creation of Exodus 14–15." In *Jerusalem Studies in Egyptology*, Vol. 1, edited by Shirun-Grumach, 137–58. Wiesbaden: Harrassowitz-Verlag, 1998.

Edwards, D., and D. Middleton. "Joint Remembering: Constructing an Account of Shared Experience through Conversational Discourse." *Discourse Processes* 9 (1986): 423–59.

Encyclopaedia Judaica, "Shavuot." Vol. 14, 1319–22. Jerusalem: Keter Publishing, 1971.

Enzel, Y., G. Kadan, and Y. Eyal "Holocene Earthquakes Inferred from a Fan-Delta Sequence in the Dead Sea Graben," *Quaternary Research* 53 (2000): 34–48.

Fagan, B. *Floods, Famines, and Emperors, El Niño and the Fate of Civilizations.* New York: Basic, 1999.

Faulstich,, E., 1998, "Studies in O.T. and N.T. Chronology." In *Chronos, Kairos, Christos II: Chronological, Nativity, and Religious Studies in Memory of Ray Summers*, edited by E. Vardaman, 97–117. Macon, Ga: Mercer University Press, 1998.

Faustino, L. "Mayon Volcano and Its Eruption." *Philippine Journal of Science* 40 (1929): 1–43.

Feiler, B. *Walking the Bible: A Journey by Land through the Five Books of Moses.* San Francisco: HarperCollins, 2001.

Finegan, J. *Light from the Ancient Past: The Archeological Background of the Hebrew-Christian Religion.* Princeton: Princeton University Press, 1951.

Finkelstein, I. *The Archaeology of the Israelite Settlement.* Jerusalem: Israel Exploration Society, 1988.

Finkelstein, I. "The Emergence of the Monarchy in Israel: The Environmental and Socio-Economic Aspects." *Journal for the Study of the Old Testament* 44 (1989): 43–74.

Finkelstein, I. "The Emergence of Israel: A Phase in the Cyclic History of Canaan in the Third and Second Millennia BCE." *FNM*, 150–78.

Finkelstein, I. *Living on the Fringe.* Sheffield: Sheffield Academic Press, 1995.

Fohrer, G. *Überlieferung und Geschichte des Exodus: Eine Analyse von Ex 1–15.* Beihefte zur Zeitschrift für die alttestamentliche Wissenschaft, no. 91. Berlin: Verlag Alfred Töpelmann, 1964.

Foucault, R., and D.-J. Stanley. "Late Quaternary Palaeoclimatic Oscillations in East Africa Recorded by Heavy Minerals in the Nile Delta." *Nature* 339 (1989): 44–46.

Fox, E. *Genesis and Exodus: A New English Rendition with Commentary and Notes.* New York: Schocken Books, 1990.

Francis, P. *Volcanoes: A Planetary Perspective.* Oxford: Clarendon Press, 1993.

Frankel, R. "Upper Galilee in the Late Bronze–Iron I Transition." *FNM*, 18–34.

Friedman, R. *Who Wrote the Bible?* New York: Summit, 1987.

Friedrich, W. *Fire in the Sea.* Cambridge: Cambridge University Press, 2000.

Friedrich, W., B. Kromer, M. Friedrich, J. Heinemeier, T. Pfeiffer, and S. Talamo. "Santorini Eruption Radiocarbon Dated to 1627–1600 B.C.," *Science* 312 (2006): 548.

Frumkin, A. "The Holocene History of Dead Sea Levels." In *The Dead Sea: The Lake and Its Setting*, edited by T. Niemi, Z. Ben-Avraham, and J. Gat, 237–48. New York: Oxford University Press, 1997.

Frumkin, A., and Y. Elitzur. "Historic Dead Sea Level Fluctuations Calibrated with Geological and Archaeological Evidence." *Quaternary Research* 57 (2002): 334–42.

Gal, Z. "Iron I in Lower Galilee and the Margins of the Jezreel Valley." *FNM*, 35–46.

Galanopoulos, A. "Die ägyptischen Plagen und der Auszug Israels aus geologischer Sicht." *Das Altertum* 10 (1964): 131–37.

Galanopoulos, A., and E. Bacon. *Atlantis: The Truth Behind the Legend.* Indianapolis: Bobbs-Merrill, 1969.

Gantz, T. *Early Greek Myth.* Baltimore: Johns Hopkins University Press, 1993.

Gardiner, A. "Davies' Copy of the Great Speos Artemidos Inscription." *JEA* 32 (1946): 43–56.

Garfunkel, Z., and Z. Ben-Avraham. "The Structure of the Dead Sea Basin," *Tectonophysics* 266 (1996): 155–76.

Garstang, J., and J. B. E. Garstang. *The Story of Jericho.* London: Hodder & Stoughton, 1940.

Gasse, F., and E. Van Campo. "Abrupt Post-Glacial Climate Events in West Asia and North Africa Monsoon Domains." *Earth and Planetary Science Letters* 126 (1994): 435–56.

Giveon, R. *Les Bédouins Shosou des Documents Égyptiens.* Leiden: Brill, 1971.

Goedicke, H. translator. "Hatshepsut's Temple Inscription at Speo Artemidos." *BAR* 7 (September–October 1981): 49.

Goedicke, H. "The Canaanite Illness," In *Festschrift Wolfgang Helk,* edited by H. Altenmüller and D. Wildung, 91–105. Studien zur Altägyptischen Kultur 11. Hamburg: Helmut Buske, 1984.

Goedicke, H. "The End of the Hyksos in Egypt." In *Egyptological Studies in Honor of Richard A. Parker,* edited by L. Lesko, 37–47. Hanover, N.H.: Brown University Press, 1986.

Goedicke, H., "The Chronology of the Thera/Santorin Explosion." *Ägypten und Levant* 3 (1992): 57–62.

Goyon, G. "Les Travaux de Chou et les Tribulations de Geb." *Kêmi* 6 (1936): 1–42.

Graesser, C. "Standing Stones in Ancient Palestine." *Biblical Archaeologist* 35 (1972): 34–63.

Gray, J. *Archaeology and the Old Testament World.* New York: Harper & Row, 1962.

Greenberg, M. "The Thematic Unity of Exodus iii–xi." In *Fourth World Congress on Jewish Studies,* Vol. 1, 51–54. Jerusalem: World Union of Jewish Studies, 1967.

Greene, M. T. *Natural Knowledge in Preclassical Antiquity.* Baltimore: Johns Hopkins University Press, 1992.

Gregg, C. *Plague: An Ancient Disease in the Twentieth Century.* Rev. ed. Albuquerque: University of New Mexico Press, 1985.

Griffith, F. "The Antiquities of Tell el Yahudiyeh." In *Egypt Exploration Fund Memoir 7.* London: Kegan Paul, 1890.

Grove, J. *The Little Ice Age.* London: Methuen, 1988.

Grüss, J. "*Saccharomyces winlocki,* die Hefe aus den Pharaonengräbern." *Tageszeitung für Brauerei* 27 (1929): 275–78.

Grüss, J. "Untersuchung von Broten aus der ägyptischen Sammlung der Staatlichen Museen zu Berlin." *Zeitschrift für ägyptische Sprache und Altertumskunde* 68 (1932): 79–80.

Guichard, F., S. Carey, M. Arthur, H. Sigurdsson, and M. Arnold. "Tephra from the Minoan Eruption of Santorini in Sediments of the Black Sea." *Nature* 363 (1993): 610–12.

Hadingham, E. "Did a Tsunami Wipe Out the Cradle of Western Civilization?" *Discovery* 29 (January 2008): 8–14.

Halpern, B. *The Emergence of Israel in Canaan*. Society of Biblical Literature Monograph no. 29. Chico, Calif.: Scholars Press, 1983.

Halpern, B. "Radical Exodus Redating Fatally Flawed." *BAR* 13 (November–December 1987): 56–61.

Halpern, B. *David's Secret Demons: Messiah, Murderer, Traitor, King*. Grand Rapids: Eerdmans, 2001.

Hardy, D., and A. Renfrew, eds. *Thera and the Aegean World III*. Vol. 3, *Chronology*. London: Thera Foundation, 1990.

Har-el, M. *The Sinai Journeys: The Route of the Exodus*. Los Angeles: Ridgefield, 1983.

Harris, J., and K. Weeks. *X-Raying the Pharaohs*. New York: Scribners, 1973.

Harris, S. L. *Agents of Chaos: Earthquakes, Volcanoes, and Other Natural Disasters*. Missoula: Mountain Press, 1990.

Hass, J., I. Richoz, W. Tinner, and L. Wick. "Synchronous Holocene Climatic Oscillations Recorded on the Swiss Plateau and at Timberline in the Alps." *The Holocene* 8 (1998): 301–09.

Hassan, F. "Nile Floods and Political Disorder in Early Egypt." *NATO*, 1–23.

Hassan, F., and S. Robinson. "High-Precision Radiocarbon Chronometry of Ancient Egypt, and Comparisons with Nubia, Palestine, and Mesopotamia." *Antiquity* 61 (1987), 119–35.

Hay, L. "What Really Happened at the Sea of Reeds?" *JBL* 83 (1964): 397–403.

Hayes, C. "Golden Calf Stories: The Relationship of Exodus 32 and Deuteronomy 9–10." In *The Idea of Biblical Interpretation: Essays in Honor of James L. Kugel*, edited by H. Nazman and J. Newman, 45–93. Leiden: Brill, 2004.

Henige, D. *Oral Historiography*. London: Longman, 1982.

Heiken, G., and F. McCoy. "Caldera Development during the Minoan Eruption, Thira, Cyclades, Greece." *JGR* 89 (1984): 8441–62.

Heiken, G., and F. McCoy. "Precursory Activity to the Minoan Eruption, Thera, Greece." *Thera III-2*, 79–88.

Hendrickson, R., *More Cunning Than Man: A Social History of Rats and Men*. New York: Stein and Day, 1983.

Hendrix, R. "A Literary Structural Analysis of the Golden-Calf Episode in Ex 32:1–33:6." *Andrews University Seminary Studies* 28 (1990): 211–17.

Herrmann, S. *A History of Israel in Old Testament Times*. Philadelphia: Fortress, 1981.

Herzog, C., and M. Gichon. *Battles of the Bible*. London: Greenhill, 1997.

Herzog, M., H.-F. Graf, C. Textor, and J. M. Oberhuber. "The Effect of Phase Changes of Water on the Development of Volcanic Plumes." *JVGR* 87 (1998): 55–74.

Higginbotham, C. *Egyptianization and Elite Emulation in Ramesside Palestine. Culture and History of the Ancient Near East*, Vol. 2. Leiden: Brill, 2000.

Higham, T. "The Experimental Study of the Transmission of Rumour." *British Journal of Psychology* 42 (1951): 42–55.

Hirst, W., and D. Manier. "Remembering as Communication: A Family Recounts Its Past." In *Remembering Our Past: Studies in Autobiographical Memory*, edited by D. Rubin, 271–90. Cambridge: Cambridge University Press, 1996.

Hoffmeier, J. "Reconsidering Egypt's Part in the Termination of the Middle Bronze Age in Palestine." *Levant* 21 (1989): 181–93.

Hoffmeier, J. *Israel in Egypt: The Evidence for the Authenticity of the Exodus Tradition.* New York: Oxford University Press, 1997.

Hoffmeier, J. *Ancient Israel in Sinai: The Evidence for the Authenticity of the Wilderness Tradition.* New York: Oxford University Press, 2005.

Holladay, J. "The Eastern Nile Delta during the Hyksos and Pre-Hyksos Periods: Toward a Systemic/Socioeconomic Understanding." *Hyksos*, 183–252.

Hort, G. 1955, "Musil, Madian, and the Mountain of the Law." In *Jewish Studies: Essays in Honour of the Very Reverend Dr. Gustav Sicher, Chief Rabbi of Prague*, 81–93. Prague: Council of Jewish Religious Communities, 1955.

Hort, G. "The Plagues of Egypt." *Zeitschrift für die alttestamentliche Wissenschaft* 69 (1957): 84–103.

Hort, G. "The Plagues of Egypt," *Zeitschrift für die alttestamentliche Wissenschaft* 70 (1958): 48–59.

Hort, G. "The Death of Qorah." *Australian Biblical Review* 7 (1959): 2–26.

Howarth, D. *Waterloo: Day of Battle.* New York: Galahad, 1968.

Hunt, J., and P. Hill. "Tephrological Implications of Beam Size–Sample-Size Effects in Electron Microprobe Analysis of Glass Shards." *Journal of Quaternary Science* 16 (2001): 105–17.

Ilan, D. "The Dawn of Internationalism—The Middle Bronze Age." *ASHL*, 297–319.

Irvin, D. "The Joseph and Moses Stories as Narrative in the Light of Ancient Near Eastern Narrative." *IJH* 180–202.

Issar, A., and M. Zohar. *Climate Change: Environment and Civilization in the Middle East.* Berlin: Springer, 2004.

Jack, J. *The Date of the Exodus: In the Light of External Evidence.* Edinburgh: T & T Clark, 1925.

Jolly, D., R. Bonnefille, and M. Roux. "Numerical Interpretation of a High Resolution Holocene Pollen Record from Burundi." *Palaeogeography, Palaeoclimatology, Palaeoecology* 109 (1994): 357–70.

Kassin, S., P. Ellsworth, and V. Smith. "The 'General Acceptance' of Psychological Research on Eyewitness Testimony: A Survey of the Experts." *American Psychologist* 44 (1989): 1089–98.

Kastens, K., and M. Cita. "Tsunami-Induced Sediment Transport in the Abyssal Mediterranean Sea." *GSABull* 92 (1981): 845–57.

Keegan, J. *The Face of Battle.* New York: Viking, 1976.

Keller, E., and N. Pinter. *Active Tectonics: Earthquakes, Uplift, and Landscape.* Upper Saddle River, N.J.: Prentice-Hall, 1996.

Keller, J. "Prehistoric Pumice Tephra on Aegean Islands." In *Thera and the Aegean World II*, edited by C. Doumas, 49–56. London: Thera Foundation, 1980.

Keller, J., W. Ryan, D. Ninkovich, and R. Altherr. "Explosive Volcanic Activity in the Mediterranean over the Past 200,000 Yr as Recorded in Deep-Sea Sediments." *GSABull* 89 (1978): 591–604.

Kemp, B. *Ancient Egypt: Anatomy of a Civilization.* London: Routledge, 1989.

Kempinski, A. "Some Observations on the Hyksos (XVth) Dynasty and Its Canaanite Origins." In *Pharaonic Egypt: The Bible and Christianity*, edited by S. Israelit-Groll, 129–37. Jerusalem: Magnes Press, 1985.

Kempinski, A. "Jacob in History," *BAR* 14 (January–February 1988): 42–47, 67.

Kenyon, K. "Some Notes on the History of Jericho in the Second Millennium B.C." *PEQ* 83 (1951): 101–38

Kenyon, K. *Digging Up Jericho*. London: E. Benn, 1957.

Kenyon, K. *Excavations at Jericho*. Vol. 1. *The Tombs Excavated in 1952–4*. London: British School of Archaeology in Jerusalem, 1960.

Kenyon, K. "Jericho." In *Archaeology and Old Testament Study*, edited by D. Thomas, 264–75. Oxford: Clarendon, 1967.

Kenyon, K. *Archaeology in the Holy Land*. 3rd ed. New York: Praeger, 1970.

Kenyon, K. "Jericho." In *Encyclopedia of Archaeological Excavations in the Holy Land*, edited by M. Avi-Yonah. Vol. 2, 550–64. Englewood Cliffs, N.J.: Prentice-Hall, 1976.

Kenyon, K., and T. Holland. *Excavations at Jericho*, Vols. 3, 4, and 5. London: British School of Archaeology in Jerusalem, 1981–83.

Keyes, D. *Catastrophe: An Investigation into the Origins of the Modern World*. New York: Ballantine, 1999.

King, P. "Circumcision—Who Did it, Who Didn't and Why." *BAR* 12 (July–August 2006): 48–55.

King, P., and L. Stager. *Life in Biblical Israel*. Louisville: Westminster/John Knox, 2001.

Kirsch, J. *Moses: A Life*. New York: Ballantine, 1998.

Kitchen, K. "The Egyptian Evidence on Ancient Jordan." In *Early Edom and Moab: The Beginning of the Iron Age in Southern Jordan*, edited by P. Bienkowski, 21–34. Sheffield: Collis, 1992.

Kitchen, K. "The Historical Chronology of Ancient Egypt: A Current Assessment." *Acta Archaeologia* 67 (1996): 1–14.

Kitchen, K. "Egyptians and Hebrews from Ra'amses to Jericho." In *The Origin of Early Israel: Current Debate, Biblical, Historical, and Archaeological Perspectives*, edited by S. Ahituv and E. Oren, 65–131. Beersheva: Ben-Gurion University of the Negev Press, 1998.

Kittleman, L. "Geologic Methods in Studies of Quaternary Tephra," In *Volcanic Activity and Human Ecology*, edited by P. Sheets and D. Grayson, 49–82. New York: Academic Press, 1979.

Kœnig, J. "La localisation du Sinaï et les traditions des scribes." *Révue d'Histoire et de Philosophie religieuses* 44 (1964): 200–35.

Kœnig, J. "Itinéraires Sinaitiques en Arabie." *Révue de l'Histoire des Religions* 166 (1964): 121–41.

Kœnig, J. *Le site de Al-Jaw dans l'Ancien Pays de Madian*. Paris: Geuthner, 1971.

Krahmalkov, C. "Exodus Itinerary Confirmed by Egyptian Evidence." *BAR* 20 (September–October 1994): 55–62, 79.

Krogman, W., and M. Baer. "Age at Death of Pharaohs of the New Kingdom, Determined by X-Ray Films." In *An X-Ray Atlas of the Royal Mummies*, edited by J. Harris and E. Wente, 188–212. Chicago: University of Chicago Press, 1980.

Krom, M., D.-J. Stanley, R. Cliff, and J. Woodward. "Nile River Sediment Fluctuations over the Past 7000 Yr and Their Key Roll in Sapropel Development." *Geology* 30 (2002): 71–74.

LaBianca, Ø., and R. Younker. "The Kingdoms of Ammon, Moab, and Edom: The Archaeology of Society in Late Bronze Iron Age Transjordan (ca. 1400–500 BCE)." *ASHL*, 399–411.

Lamb, H. *The Crusades*. New York: Bantam, 1962.

Lee, H. translator. *Aristotle Meterologica* Cambridge: Harvard University Press, 1952.

Lehner, M. "The Giza Plateau Mapping Project: 1993–94 Annual Report." www-oi.uchicago.edu/OR/AR/93–94/93–94_Giza.html.

Lenes, J., B. Darrow, C. Cattrall, C. Heil, M. Callahan, G. Vargo, R. Byrne, et al. "Iron Fertilization and the *Trichodesmium* Response on the West Florida Shelf." *Limnology and Oceanography* 46 (2001): 1261–77.

Lichtheim, M. *Ancient Egyptian Literature*. Vol. 1, *The Old and Middle Kingdoms*. Berkeley: University of California Press, 1973.

Lichtheim, M. *Ancient Egyptian Literature*. Vol. 2, *The New Kingdom*. Berkeley, University of California Press, 1976.

Linton, M. "Transformations of Memory in Everyday Life." In *Memory Observed: Remembering in Natural Contexts*, edited by U. Neisser, 77–91. San Francisco: Freeman, 1982.

Linton, R. "Nativistic Movements." *American Anthropologist* 45: (1943): 230–40.

Liritzis, I., C. Michael, and R. Galloway. "A Significant Aegean Volcanic Eruption during the Second Millennium B.C. Revealed by Thermoluminescence Dating." *Geoarchaeology* 11 (1996): 361–71.

Livingston, D. "Further Considerations on the Location of Bethel at El-Bireh." *PEQ* 126 (1994): 154–59.

Loftus, E., and W. Marburger, "Since the Eruption of Mt. St. Helens, Has Anyone Beaten You Up? Improving the Accuracy of Retrospective Reports with Landmark Events." *Memory and Cognition* 11 (1983): 114–20.

Lockwood, J. *World Climatology: An Environmental Approach*. New York: St. Martin's Press, 1974.

Lord, A. *The Singer of Tales*. Cambridge: Harvard University Press, 1960.

Lucas, A. *The Route of the Exodus of the Israelites from Egypt*. London: Edward Arnold, 1938.

Magny, M. "Solar Influences on Holocene Climatic Changes Illustrated by Correlations between Past Lake-Level Fluctuations and the Atmospheric ^{14}C Record." *Quaternary Research* 40 (1993): 1–9.

Magny, M. "Successive Oceanic and Solar Forcing Indicated by Younger Dryas and Early Holocene Climatic Oscillations in the Jura." *Quaternary Research* 43 (1995): 279–85. Malamat, A. "Last Kings of Judah and Fall of Jerusalem," *IEJ* 18 (1968): 137–55.

Malamat, A., "The Proto-History of Israel: A Study in Method." In *The Word of the Lord Shall Go Forth: Essays in Honor of David Noel Freedman* edited by C. Meyers and M. O'Connor, 303–13. Winona Lake, Ind.: Eisenbrauns, 1983.

Malamat, A. "The Exodus: Egyptian Analogies." *Exodus*, 15–26.

Maley, J. "Middle to Late Holocene Changes in Tropical Africa and Other Continents: Paleomonsoon and Sea Surface Temperature Variations." *NATO*, 611–40.

Manning, S. *A Test of Time: The Volcano of Thera and the Chronology and History of the Aegean and East Mediterranean in the Mid Second Millennium BC*. Oxford: Oxbow, 1999.

Manning, S., B. Kromer, P. Kuniholm, and M. Newton, "Anatolian Tree-Rings and a New Chronology for the East Mediterranean Bronze-Iron Ages." *Science* 294 (2001): 2532–35.

Manning, S., B. Kromer, P. Kuniholm, and M. Newton. "Confirmation of Near-Absolute Dating of East Mediterranean Bronze-Iron Dendrochronology." *Antiquity* 2003. http://antiquity.ac.uk/ProjGall/Manning/manning.html.

Manning, S., C. Ramsey, W. Kutschera, T. Higham, B. Kromer, P. Steier, and E. Wild. "Chronology for the Aegean Late Bronze Age 1700–1400 B.C." *Science* 312 (2006): 565–69.

Marcolongo, B. "Évolution du paléo-environment dans la partie orientale du Delta du Nil depuis la transgression flandrienne (8,000 B.P.) par rapport aux modèles de peuplement anciens." In *Sociétés Urbaines en Égypte et au Soudan*, 23–31. Cahiers de Recherches de l'Institut de Papyrologie et d'Égyptologie de Lille 14, Lille, 1992.

Martin, C. "The Recent Eruption of Katmai Volcano in Alaska." *National Geographic* 24, no. 2 (1913): 131–81.

Mazar, A. *Archaeology of the Land of the Bible—10,000–586 B.C.E.* New York: Doubleday, 1992.

McCarter, P. "The Patriarchal Age: Abraham, Isaac and Jacob." In *Ancient Israel: A Short History from Abraham to the Roman Destruction of the Temple*, edited by H. Shanks, 1–29. Washington: Biblical Archaeology Society, 1988.

McCarthy, D. "Moses' Dealings with Pharaoh: Ex 7,8–10, 27." *Catholic Biblical Quarterly* 27 (1965): 336–45.

McCoy, F. "Areal Distribution, Redeposition and Mixing of Tephra within Deep-Sea Sediments of the Eastern Mediterranean Sea." In *Tephra Studies*, edited by S. Self and R. Sparks, 245–54. Dordrecht: Reidel, 1981.

McCoy, F., and G. Heiken. "The Late-Bronze Age Explosive Eruption of Thera (Santorini), Greece: Regional and Local Effects." In *Volcanic Hazards and Disasters in Human Antiquity*, edited by F. McCoy and G. Heiken, 43–70. Geological Society of America Special Paper no. 345, Boulder, Co.: 2000.

McCoy, F., and G. Heiken. "Tsunami Generated by the Late Bronze Age Eruption of Thera (Santorini) Greece." *Pure and Applied Geophysics* 157 (2000): 1227–56.

McNeill, W. 1977, *Plagues and Peoples*. Garden City, NY: Doubleday, 1977.

Meek, T. "Moses and the Levites." *American Journal of Semitic Languages and Literatures* 56 (1939): 113–20.

Mennis, M. "The Existence of Yomba Island near Madang: Fact or Fiction?" *Oral History* 6, (1978): 2–81.

Mercer, S. *The Religion of Ancient Egypt*. London: Luzac, 1949.

Meskhidze, M., W. Chameides, and A. Nenes. "Dust and Pollution: A Recipe for Enhanced Ocean Fertilization." *JGR* 110 (2005), D03301.

Michael, H. "Radio-carbon Dates from Akrotiri on Thera." *Temple University Aegean Symposium, First*, edited by P. Betancourt, 7–9. Philadelphia: Temple University Press, 1976.

Michaud, V., R. Clocchiatti, and S. Sbrana. "The Minoan and post-Minoan eruptions, Santorini (Greece), in the Light of Melt Inclusions: Chlorine and Sulphur Behavior." *JVGR* 99 (2000): 195–214.

Miller, J. C. "Introduction: Listening for the African Past." In *The African Past Speaks: Essays on Oral Tradition and History*, edited by J. C. Miller, 1–59. Hamden, Conn.: Archon, 1980.

Miller, J. M. "The Israelite Occupation of Canaan." *IJH*, 213–84.

Miller, J. M. "The Israelite Journey through (around) Moab and Moabite Toponymy" *JBL* 108 (1989): 577–95.

Miller, J. M. "Early Monarchy in Moab?" In *Early Edom and Moab, The Beginning of the Iron Age in Southern Jordan*, edited by P. Bienkowski, 77–92. Sheffield: Collis, 1992.

Minoura, K., G. Papadopoulos, T. Takahashi, A. Yalciner, F. Imanura, U. Kuran, and T. Nakamura. "Discovery of Minoan Tsunami Deposits." *Geology* 28 (2000): 59–62.

Moldenke, H., and A. Moldenke. *Plants of the Bible*. Waltham: Chronica Botanica, 1952.

Moodie, D., A. Catchpole, and K. Abel. "Northern Athapascan Oral Traditions and the White River Volcano." *Ethnohistory* 39 (1992): 148–71.

Morris, J. *The Age of Arthur: A History of the British Isles from 350 to 650*. New York: Scribner's, 1973.

Morrison, D. "Ancient Egypt Cities Leveled by Massive Volcano, Lava Find Suggests." *National Geographic News* (April 2, 2007). http://news.nationalgeographic.com/news/pf/36949218.html.

Murnane, W. *Ancient Egyptian Coregencies*. Studies in Ancient Oriental Civilization 40. Chicago: Oriental Institute, 1977.

Musil, A. "Vorbericht über seine letzie Reise nach Arabien." *Anzerger der Kaiserlichen Akademie der Wissenschaften in Wein, Philosophisch-historische klasse* 48 (1911): 139–59.

Musil, A. *The Northern Heğâz*. New York: American Geographical Society, 1926.

Myres, J. *Who Were the Greeks?* New York: Biblo and Tannen, 1967.

Na'aman, N. "Habiru and Hebrews: The Transfer of a Social Term to the Literary Sphere." *JNES* 45 (1986): 271–88.

Na'aman, N. "The Hurrians and the End of the Middle Bronze Age in Palestine." *Levant* 26 (1994): 175–87.

Na'aman, N. "The 'Conquest of Canaan' in the Book of Joshua and in History." *FNM*, 218–81.

Nagy, G. *Greek Mythology and Poetics*. Ithaca, N.Y.: Cornell University Press, 1990.

Najera, J., B. Liese, and J. Hammer. *Malaria: New Patterns and Perspectives*. World Bank Technical Paper no. 183. Washington, 1992.

Nania, J., and T. Bruha. "In the Wake of Mount St Helens." *Annals of Emergency Medicine* 11 (1982): 184–91.

Narcisi, B., and L. Vezzoli. "Quaternary Stratigraphy of Distal Tephra Layers in the Mediterranean—an Overview." *Global and Planetary Change* 21 (1999): 31–50.

National Institutes of Mental Health, *Training Manual for Human Service Workers in Major Disasters*. U.S. Dept. of Health, Education and Welfare Publication No. (ADM) 77–538, 1978.

Neel, J. "The Population Structure of an Amerindian Tribe, the Yamomana." *Annual Review of Genetics* 12 (1979): 365–413.

Neev, D. "The Pelusium Line: A Major Transcontinental Shear." *Tectonophysics* 38 (1977): T1–T8.

Neev, D, N. Bakler, and K. Emery. *Mediterranean Coasts of Israel and Sinai*. London: Taylor & Francis, 1987.

Neev, D., and K. Emery. *The Destruction of Sodom, Gomorrah, and Jericho: Geological, Climatological, and Archaeological Background*. Oxford: Oxford University Press, 1995.

Neev, D., and G. Friedman. "Late Holocene Tectonic Activity along the Margins of the Sinai Subplate." *Science* 202 (1978): 427–29.

Neev, D., J. Hall, and J. Saul. 1972, "The Pelusium Megashear System across Africa and Associated Lineament Swarms." *JGR* 87 (1972): 1015–30.

Neufeld, E. "Insects as Warfare Agents in the Ancient Near East (Ex. 23:28; Deut. 7:20; Josh.24:12; Isa. 7:18–20)," *Orientalia* 49 (1980): 30–57.

Neumann, J. "Climatic Changes in Europe and the Near East in the Second Millennium BC." *Climatic Change* 23 (1993): 231–45.

Neumann, J., and S. Parpola. "Climatic Change and the Eleventh-Tenth-Century Eclipse of Assyria and Babylonia." *JNES* 46 (1987): 161–82.

Newberry, P. *Beni Hasan*, Vol. 1. London: Archaeological Survey of Egypt and K. Paul, Trench, Trübner, 1893.

Newby, P. *Warrior Pharaohs: The Rise and Fall of the Egyptian Empire*. London: Faber and Faber, 1980.

Nicholson, S. 1980, "Saharan Climates in Historic Times." In *The Sahara and the Nile: Quaternary Environments and Prehistoric Occupation in Northern Africa*, edited by M. Williams and H. Faure, 173–200. Rotterdam: Balkema, 1980.

Niditch, S. *Oral World and Written Word*. Louisville: Westminster/John Knox, 1996.

Ninkovich, D., and B. Heezen. "Santorini Tephra." In *Submarine Geology and Geophysics*, edited by W. Wittard and R. Bradshaw, 415–53. London: Butterworths, 1965.

Noth, M. "Der Wallfahrtsweg zum Sinai." *Palästinafahrbuch* 36 (1940): 5–28.

Noth, M. *Exodus: A Commentary*. Philadelphia: Westminster, 1962.

Nur, A., and E. Cline, "Poseidon's Horses: Plate Tectonics and Earthquake Storms in the Late Bronze Age Aegean and Eastern Mediterranean." *Journal of Archaeological Science* 27 (2000): 43–63.

Nur, A., and E. Cline. "What Triggered the Collapse? Earthquake Storms." *Archaeology Odyssey* 4 (September–October 2001): 30–36, 62.

Ofer, A. "'All the Hill Country of Judah': From a Settlement Fringe to a Prosperous Monarchy." *FNM*, 92–121.

Oren, E. "'Governors' Residencies' in Canaan under the New Kingdom: A Case Study of Egyptian Administration." *Journal of the Society for the Study of Egyptian Antiquities* 14 (1984): 37–56.

Oren, E. "The 'Ways of Horus' in North Sinai." In *Egypt, Israel, Sinai: Archaeological and Historical Relationships in the Biblical Period*, edited by A. Rainey, 69–119. Tel Aviv: Tel Aviv University, 1987.

Oren, E. "The 'Kingdom of Sharuhen' and the Hyksos Kingdom." *Hyksos*, 253–84.

Oskarsson, N. "The Interaction between Volcanic Gases and Tephra: Fluorine Adhering to Tephra of the 1970 Hekla Eruption." *JVGR* 8 (1980): 250–66.

Parr, P. "Edom and the Hejaz." In *Early Edom and Moab: The Beginning of the Iron Age in Southern Jordan*, edited by P. Bienkowski, 41–46. Sheffield: Collis, 1992.

Pearce, N., W. Eastwood, J. Westgate, and W. Perkins. "Trace-element Composition of Single Glass Shards in Distal Minoan Tephra from SW Turkey." *Journal of the Geological Society of London* 159 (2002): 545–56.

Pearce, N., J. Westgate, S. Preece, W. Eastwood, and W. Perkins. "Identification of Aniakchak (Alaska) tephra in Greenland ice core challenges the 1645 BC date for Minoan eruption of Santorini." *Geochemistry Geophysics Geosystems* 5 (2004): Q03005, doi:10.1029/2003GC000672.

Pennington, R. "Causes of Early Human Population Growth." *American Journal of Physical Anthropology* 99 (1996): 259–74.

Perfect, T., T. Hollins, and A. Hunt. "Practice and Feedback Effects on the Confidence-Accuracy Relation in Eyewitness Memory." *Memory* 8 (2000): 235–44.

Perry, R., and J. Fetherston. "*Yersinia pestis*—Etiologic Agent of Plague." *Clinical Microbiology Reviews* 10 (1997): 35–66.

Pétrequin, P. "Management of Architectural Woods and Variations in Population Density in the Fourth and Third Millennia B.C. (Lakes Chalain and Clairvaux, Jura, France)." *Journal of Anthropological Archaeology* 15 (1996):1–19.

Petrie, W. *Hyksos and Israelite Cities*. British School of Archaeology in Egypt Publication no. 12. London, 1906.

Philby, H. *The Land of Midian*. London: Ernest Benn, 1957.

Pickrell, J. "Aerial War Against Disease: Satellite Tracking of Epidemics is Soaring," *Science News* 161 (April 6, 2002): 218–20.

Pirenne, J. "Le Site Préislamique de Al-Jaw, la Bible, le Coran et le Midrash," *Révue Biblique* 82 (1975): 34–69.

Pielou, E. *The Energy of Nature*. Chicago: University of Chicago Press, 2001.

Poland, J., T. Quan, and A. Barnes. 1994, "Plague." In *Handbook of Zoonoses, Section A: Bacterial, Rickettsial, Chlamydial, and Mycotic*, 2nd ed., edited by G. Beran, 93–112. Boca Raton, Fl.: CRC Press, 1994.

Portugali, J. 1994, "Theoretical Speculations on the Transition from Nomadism to Monarchy." *FNM*, 203–17.

Prag, K. "The Intermediate Early Bronze–Middle Bronze Age Sequences at Jericho and Tell Iktanu Reviewed." *BASOR* 264 (1986): 61–72.

Prager, E. *Furious Earth: The Science and Nature of Earthquakes, Volcanoes, and Tsunamis*. New York: McGraw-Hill, 2000.

Price, C., L. Stone, A. Huppert, B. Rajagopalan, and P. Alpert. "A Possible Link between El Niño and Precipitation in Israel." *Geophysical Research Letters* 25 (1998): 3963–66.

Pritchard, J., ed. *Ancient Near Eastern Texts Relating to the Old Testament*. Princeton: Princeton University Press, 1969.

Pyle, D. "New Estimates for the Volume of the Minoan Eruption." *Thera III-2*, 113–21.

Rainey, A. "Review of Bimson 1978." *IEJ* 30 (1980): 249–51.

Rainey, A. "The Biblical Shephelah of Judah." *BASOR* 251 (1983): 1–22.

Read, J., D. Lindsay, and T. Nicholls. "The Relation between Confidence and Accuracy in Eyewitness Identification Studies: Is the Conclusion Changing?" In *Eyewitness Memory: Theoretical and Applied Perspectives*, edited by C. Thompson, D. Herrman, J. Read, D. Bruce, D. Payne, and M. Toglia, 107–30. Mahwah, N.J.: Lawrence Erlbaum, 1998.

Redford, D. "Exodus I 11." *Vetus Testamentum* 13 (1963): 401–18.

Redford, D. "The Coregency of Tuthmosis III and Amenophis II." *JEA* 51 (1965): 107–22.

Redford, D. 1987, "An Egyptological Perspective on the Exodus Narrative." In *Egypt, Israel, Sinai: Archaeological and Historical Relationships in the Biblical Period*, edited by A. Rainey, 144–45. Tel Aviv: Tel Aviv University Press, 1987.

Redford, D. *Egypt, Canaan, and Israel in Ancient Times*. Princeton: Princeton University Press, 1992.

Redford, D. "Observations on the Sojourn of the Bene-Israel." *Exodus*, 57–66.

Redford, D. "Textual Sources for the Hyksos Period." *Hyksos*, 1–44.

Redmount, C. *On an Egyptian/Asiatic Frontier: An Archaeological History of the Wadi Tumilat*. 4 Vols. Chicago: The University of Chicago Department of Near Eastern Languages and Civilizations Ph.D. Dissertation, 1989.

Redmount, C. "Pots and Peoples in the Egyptian Delta: Tell El-Maskhuta and the Hyksos." *Journal of Mediterranean Archaeology* 8, no. 2 (1995): 61–89.

Ribeiro, R. "Brazilian Messianic Movements." In *Millennial Dreams in Action*, edited by S. L. Thrupp, 55–69. The Hague: Mouton, 1962.

Robock, A. "Volcanic Eruptions and Climate." *Reviews of Geophysics* 38 (2000): 191–219.

Rose, M. "Mystery Mummy." *Archaeology* 56, (2003): 18–25.

Rose, W., Jr., G. Bluth, and G. Ernst. "Integrating Retrievals of Volcanic Cloud Characteristics from Satellite Remote Sensors: A Summary." *Philosophical Transactions of the Royal Society (London)*, 358A (2000): 1585–1606.

Rose, W., Jr., G. Bluth, D. Schneider, G. Ernst, C. Riley, L. Henderson, and R. McGimsey. "Observations of Volcanic Clouds in Their First Few Days of Atmospheric Residence: The 1992 Eruption at Crater Peak, Mount Spurr Volcano, Alaska." *Journal of Geology* 109 (2001): 677–95.

Rose, W., Jr., D. Delene, D. Schneider, G. Bluth, A. Krueger, I. Sprod, C. McKee, H. Davies, and G. Ernst. "Ice in the 1994 Rabaul Eruption Cloud: Implications for Volcano Hazard and Atmospheric Effects." *Nature* 375 (1995): 477–79.

Rose, W., Jr., C. Riley, and S. Dartevelle. "Sizes and Shapes of 10-Ma Distal Fall Pyroclasts in the Ogallala Group, Nebraska." *Journal of Geology* 111 (2003): 115–24.

Rosen, A. "Environmental Change and Settlement at Tel Lachish, Israel." *BASOR* 263 (1986): 55–60.

Rothlisberger, F. *10 000 Jahre Gletschergeschichte der Erde*. Frankfurt am Main and Salzburg: Aarau and Sauerländer, 1986.

Round, F. *The Ecology of Algae*. Cambridge: Cambridge University Press, 1981.

Rumney, G. *Climatology and the World's Climates*. New York: Macmillan, 1968.

Ryholt, K. *The Political Situation in Egypt during the Second Intermediate Period, c. 1800–1550 B.C.* Carsten Niebuhr Institute Publication no. 20. Copenhagen: Museum Tusculanum Press, 1997.

Said, R. *The River Nile: Geology, Hydrology, and Utilization*. New York: Pergamon, 1994.

Sampson, A. "Minoan Finds from Tilos," *Athens Annals of Archaeology* 13 (1981): 68–73.

Sampson, A., and I. Liritzis. "Archaeological and Archaeometrical Research at Yali, Nissiros." *TÜBA-AR* 2 (1998): 101–15.

Sarris, P. "The Justinianic Plague: Origins and Effects."*Continuity and Change* 17 (2002): 169–82.

Sauneron, S. *The Priests of Ancient Egypt*. Ithaca, N.Y.: Cornell University Press, 2000.

Säve-Söderbergh, T. *The Navy of the Eighteenth Egyptian Dynasty*. Uppsala: A.-B. Lundequistska, 1946.

Scarth, A. *Volcanoes, An Introduction*. College Station: Texas A&M University Press, 1994.

Scarth, A. *Vulcan's Fury: Man Against the Volcano*. New Haven: Yale University Press, 1999.

Scarth, A., and J.-C. Tanguy. *Volcanoes of Europe*. Oxford: Oxford University Press, 2001.

Schwartz, B. "What Really Happened at Mount Sinai." *Bible Review* 13 (1997): 20–30, 46.

Scott, S., and C. Duncan. *Biology of Plagues: Evidence from Historical Populations*. Cambridge: Cambridge University Press, 2001.

Sellin, E., and C. Watzinger. *Jericho: die Ergebnisse der Ausgrabungen*. Leipzig: J. C. Hinrichs, 1913.

Sestini, G. "Nile Delta: A Review of Depositional Environments and Geological History." In *Deltas: Sites and Traps for Fossil Fuels*, edited by M. Whateley and K. Pickering, 99–127. Geological Society Special Publication no. 41. London, 1989.

Shafei, Bey, A. "Historical Notes on the Pelusiac Branch, the Red Sea Canal and the Route of the Exodus." *Bulletin de la Société Royale de Géographie d'Égypte* 21 (1946): 231–87.

Shanks, H. "The Exodus and the Crossing of the Red Sea, According to Hans Goedicke." *BAR* 7 (September–October 1981): 42–50.

Shaw, I. "Introduction: Chronologies and Cultural Change in Egypt." In *The Oxford History of Ancient Egypt*, edited by I. Shaw, 1–15. Oxford: Oxford University Press, 2000.

Shea, W. "Exodus, Date of the." *International Standard Bible Encyclopedia*. Vol. 1, 230–38. Grand Rapids: Eerdmans, 1982.

Sheets, P. "Environmental and Cultural Effects of the Ilopango Eruption in Central America." In *Volcanic Activity and Human Ecology*, edited by P. Sheets and D. Grayson, 548–53. New York: Academic Press, 1979.

Shennan, S. "Population, Culture History, and the Dynamics of Culture Change." *Current Anthropology* 41 (2000): 1–33 of the online edition.

Siddall, M., E. Rohling, A. Almogi-Labin, Ch. Hemleben, D. Meischner, I. Schmelzer, and D. Smeed. "Sea-Level Fluctuations During the Last Glacial Cycle." *Nature* 423 (2003): 853–858.

Sigurdsson, H., S. Carey, and J. Devine. "Assessment of Mass, Dynamics and Environmental Effects of the Minoan Eruption of Santorini Volcano." *Thera III-2*, 100–12.

Sigurdsson, H., S. Carey, M. Alexandri, G. Vougioukalakis, K. Croff, C. Roman, D. Sakellariou, et al. "Marine Investigations of Greece's Santorini Volcanic Field." *Eos* 87, no. 34 (August 22, 2006): 337, 342.

Simkin, T., and L. Siebert. *Volcanoes of the World*. Tucson: Geoscience Press, 1994.

Simons, J. *Geographical and Topographical Texts of the Old Testament*. Leiden: Brill, 1959.

Singer, A. *The Soils of Israel*. Berlin: Springer, 2007.

Singer, I. "Merneptah's Campaign to Canaan and the Egyptian Occupation of the Southern Coastal Plain of Palestine in the Ramesside Period." *BASOR* 269 (1988): 1–10.

Singer, I. "Egyptians, Canaanites, and Philistines in the Period of the Emergence of Israel." *FNM*, 282–338.

Smith, H. "The story of Onchsehshonqy." *Serapis: The American Journal of Egyptology* 6 (1980): 133–66.

Smith, S. *Wretched Kush: Ethnic Identities and Boundaries in Egypt's Nubian Empire*. London: Routledge, 2003.

Smithsonian Institution. *Global Volcanism Program: Volcanic Activity Reports*. www .volcano.si.edu/gvp/volcano/region02/africa_e/nyamura/var.htm and africa_e/ nyiragon/var.htm.

Sneh, A., T. Weissbrod, and I. Perath. "Evidence for an Ancient Egyptian Frontier Canal," *American Scientist* 63 (1975): 542–48.

Soloviev, S., O. Solovieva, C. Go, K. Kim, and N. Shchetnikov. *Tsunamis in the Mediterranean Sea 2000 B.C.–2000 A.D.* Dordrecht: Kluwer, 2000.

Southon, J. "A First Step to Reconciling the GRIP and GISP2 Ice-Core Chronologies, 0–14,500 yr B.P." *Quaternary Research* 57 (2002): 32–37.

Sparks, R., S. Brazier, T. Huang, and D. Muerdter. "Sedimentology of the Minoan Deep-Sea Tephra Layer in the Aegean and Eastern Mediterranean." *Marine Geology* 54 (1983): 141–54.

Sparks, R., and T. Huang. "The Volcanological Significance of Deep-Sea Ash Layers Associated with Ignimbrites." *Geological Magazine* 117 (1980): 425–36.

Sparks, R., and C. Wilson. "The Minoan Deposits: A Review of Their Characteristics and Interpretation." *Thera III-2*, 89–99.

Spier, L. *Yuman Tribes of the Gila River*. Chicago: University of Chicago Press, 1933.

Stager, J., B. Cumming, and L. Meeker. "A High-Resolution 11,400-Yr Diatom Record from Lake Victoria, East Africa." *Quaternary Research* 47 (1997): 81–89.

Stager, L. "The Archaeology of the Family in Ancient Israel." *BASOR* 260 (1985): 1–35.

Stager, L. E. "The Song of Deborah: Why Some Tribes Answered the Call and Others Did Not." *BAR* 15 (January–February 1989): 51–64.

Stager, L. "The Impact of the Sea Peoples in Canaan (1185–1050 BCE)." *ASHL*, 332–48.

Stanley, D.-J., M. Krom, R. Cliff, and J. Woodward. "Nile Flow Failure at the End of the Old Kingdom, Egypt: Strontium Isotopic and Petrologic Evidence," *Geoarchaeology* 18 (2003): 395–402.

Stanley, D.-J., and H. Sheng. "Volcanic Shards from Santorini (Upper Minoan Ash) in the Nile Delta, Egypt." *Nature* 320 (1986): 733–35.

Stanley, D.-J., and A. Warne. "Nile Delta: Recent Geological Evolution and Human Impact." *Science* 260 (1993): 628–34.

Stephenson, G., H. Brandstätter, and W. Wagner. "An Experimental Study of Social Performance and Delay on the Testimonial Validity of Story Recall." *European Journal of Social Psychology* 13 (1983): 175–91.

Stern, E. *Archaeology of the Land of the Bible.* Vol. 2, *The Assyrian, Babylonian, and Persian Periods (732–332 B.C.E.).* New York: Doubleday, 2001.

Stern, M. *Greek and Latin Authors on Jews and Judaism.* Vol. I. Jerusalem: Magnes Press, 1974.

Stieglitz, R. "Ancient Records and the Exodus Plagues." *BAR* 13 (November–December 1987): 47–49.

Street-Perrott, F., J. Holmes, M. Waller, M. Allen, N. Barber, P. Fothergill, D. Harkness et al. "Drought and Dust Deposition in the West African Sahel: A 5500-Year Record from Kajemarum Oasis, Northeastern Nigeria." *The Holocene* 10 (2000): 293–302.

Street-Perrott, F., and R. Perrott. "Abrupt Climate Fluctuations in the Tropics: The Influence of Atlantic Ocean Circulation." *Nature* 343 (1990): 607–12.

Strothers, R. "Volcanic Dry Fogs, Climate Cooling, and Plague Pandemics in Europe and the Middle East." *Climatic Change* 42 (1999): 713–23.

Strothers, R. "Climate and Demographic Consequences of the Massive Volcanic Eruption of 1258." *Climatic Change* 45 (2000): 361–74.

Stuiver, M., P. Grootes, and T. Braziunas. "The GISP2 δ18O Climate Record of the Past 16,500 Years and the Role of the Sun, Ocean, and Volcanoes." *Quaternary Research* 44 (1995): 341–54.

Sullivan, D. "The Discovery of Santorini Minoan Tephra in Western Turkey," *Nature* 333 (1988): 552–54.

Tabazedeh, A., and R. Turco. "Stratospheric Chlorine Injection by Volcanic Eruption: HCL Scavenging and Implications for Ozone," *Science* 260 (1993): 1082–86.

Taylor, J. *Death and the Afterlife in Ancient Egypt.* London and Chicago: University of Chicago Press, 2001.

Telford, R., and H. Lamb. "Groundwater-Mediated Response to Holocene Climatic Change Recorded by the Diatom Stratigraphy of an Ethiopian Crater Lake." *Quaternary Research* 52 (1999): 63–75.

Thomas, R. *Literacy and Orality in Ancient Greece.* Cambridge: Cambridge University Press, 1992.

Thompson, C., J. Skowronski, and D. Lee. "Telescoping in Dating Naturally Occurring Events." *Memory and Cognition* 16 (1988): 461–68.

Tyldesley, J. *The Private Lives of the Pharaohs.* New York: TV Books, 2000.

Ussishkin, D. "Notes on the Fortifications of the Middle Bronze II Period at Jericho and Shechem." *BASOR* 276 (1989): 29–53.

Valbelle, D., F. Le Saout, M. Chartier-Reymond, M. Abd El-Samie, C. Traunecker, G. Wagner, J.-Y. Carrez-Maratray et al. "Reconnaissance Archéologique à la Pointe Orientale du Delta: Rapport Préliminaire sur les Saisons 1990 et 1991." In *Sociétés Urbaines en Égypte et au Soudan,* 11–22. Cahiers de Recherches de l'Institut de Papyrologie et d'Égyptologie de Lille 14. Lille: 1992.

Van Seeters, J. "The Date for the 'Admonitions' in the Second Intermediate Period." *JEA* 50 (1964): 13–23.

Van Seters, J. *The Life of Moses: The Yahwist as Historian in Exodus-Numbers.* Louisville: Westminster/John Knox, 1994.

Vansina, J. "Memory and Oral Tradition." In *The African Past Speaks: Essays on Oral Tradition and History,* edited by J. C. Miller, 262–79. Hamden, Conn.: Archon, 1980.

Vansina, J. *Oral Tradition as History*. Madison: University of Wisconsin Press, 1985.

Vitaliano, D. *Legends of the Earth: Their Geologic Origins*. Bloomington: University of Indiana Press, 1973.

Voltzinger, N., and A. Androsov. "Modeling the Hydrodynamic Situation of the Exodus." *Izvestiya, Atmospheric and Oceanic Physics* 39 (2003): 482–96.

Wacholder, B. *Eupolemus: A Study of Judaeo-Greek Literature*. Cincinnati and New York: Hebrew Union College and Jewish Institute of Religion, 1974.

Waddell, W. *Manetho*. Cambridge: Harvard University Press, 1940.

Wagenaar, W. "My Memory: A Study of Autobiographical Memory over Six Years." *Cognitive Psychology* 18 (1986): 225–52.

Wallace, A. "Revitalization Movements." *American Anthropologist* 58 (1956): 264–81.

Waltham, A. "Geological Hazards." In *The Cambridge Encyclopedia of Earth Sciences*, edited by D. Smith, 436–41. Scarborough, Ontario: Prentice-Hall Canada and Cambridge University Press, 1981.

Ward, W. "Summary and Conclusions." *Exodus*, 105–12.

Ward, W., and W. Dever. *Scarab Typology and Archaeological Context: An Essay on Middle Bronze Age Chronology*. San Antonio, Tex.: Van Siclen Books, 1994.

Watkins, N., R. Sparks, H. Sigurdsson, T. Huang, A. Federman, S. Carey, and D. Ninkovich. "Volume and Extent of the Minoan Tephra from Santorini Volcano: New Evidence from Deep-Sea Sediment Cores." *Nature* 271 (1978): 122–26.

Watson, J. "Krakatoa's Echo?" *Journal of the Polynesian Society* 72 (1963): 152–55.

Watzinger, C. "Zur Chronologie der Schichten von Jericho." *Zeitschrift der Deutschen Morgenländischen Gessellschaft* 80 (1926): 131–36.

Weinfeld, M. "Pentecost as Festival of the Giving of the Law." *Immanuel* 8, no. 1 (1978): 7–12.

Weinstein, J. "The Egyptian Empire in Palestine: A Reassessment." *BASOR* 241 (1981): 1–28.

Weinstein, J. M. "Egypt and the Middle Bronze IIC/Late Bronze IA Transition in Palestine." *Levant* 23 (1991): 105–15.

Weinstein, J. "The Chronology of Palestine in the Early Second Millennium B.C.E." *BASOR* 288 (1992): 27–46.

Weinstein, J. "Reflections on the Chronology of Tell el-Dab'a." In *Egypt, the Aegean and the Levant: Interconnections in the Second Millennium B.C.*, edited by W. Davies and L. Schofield, 84–90. London: British Museum Publications, 1995.

Weinstein, J. "Exodus and Archaeological Reality." *Exodus*, 87–103.

Weiss, B. "The Decline of Late Bronze Age Civilization as a Possible Response to Climatic Change." *Climatic Change* 4 (1982): 173–98.

Wente, E. "Genealogy of the Royal Family." In *An X-Ray Atlas of the Royal Mummies*, edited by J. Harris and E. Wente, 122–56. Chicago: University of Chicago Press, 1980.

Wente, E. "Age at Death of Pharaohs of the New Kingdom Determined from Historical Sources." In *An X-Ray Atlas of the Royal Mummies*, edited by J. Harris and E. Wente, 238–86. Chicago: University of Chicago Press, 1980.

Wente, E. "Who Was Who among the Royal Mummies." *The Oriental Institute News and Notes* 144 (1995) and www-oi.uchicago.edu/OI/IS/WENTE/NN_Win95/NN_Win95.htm (9 pages, electronic version revised July 9, 2002).

Wernsdorfer, G., and W. Wernsdorfer. "Social and Economic Aspects of Malaria and Its Control." In *Malaria: Principles and Practice of Malariology.*, Vol. 2, edited by W. Wernsdorfer and I. McGregor, 1421–71. Edinburgh: Churchill Livingstone, 1988.

Whiston, W. Translator. *The Works of Josephus*. Peabody, Mass.: Hendrickson, 1987.

White, R. "Memory for Personal Events." *Human Learning* 1 (1982): 171–83.

Wilcox, R. "Some Effects of Recent Volcanic Ash Falls with Especial Reference to Alaska." U.S. Geological Survey Bulletin 1028-N, 409–74. Washington: Government Printing Office, 1959.

Wilkinson, J. *A Popular Account of the Ancient Egyptians*, Vol. 2 . New York: Bonanza Books, 1989.

Wilson, I. *Exodus: The True Story Behind the Biblical Account*. San Francisco: Harper Row, 1985.

Wilson, R. *Genealogy and History in the Biblical World*. New Haven: Yale University Press, 1977.

Wilson, R. "Genealogy, Genealogies." In *Anchor Bible Dictonary*, Vol. 2, 929–32. New York: Doubleday, 1992.

Winchester, S. *Krakatoa, The Day the World Exploded: August 27, 1883*. New York: HarperCollins, 2003.

Winnett, F. "The Arabian Genealogies in the Book of Genesis." In *Translating and Understanding the Old Testament*, edited by H. Frank and W. Reed, 187–96. Nashville: Abingdon, 1970.

Winningham, R., I. Hyman, Jr., and D. Dinnel. "Flashbulb Memories? The Effects of When the Initial Memory Report Was Obtained," *Memory* 8 (2000): 209–16.

Wise, M. *The First Messiah*. San Francisco: HarperSanFrancisco, 1999.

Wissmann, H. von. "3,1 The Volcanoes of West Arabia." In *Catalog of Active Volcanoes of the World Including Solfatara Fields: Part 16, "Arabia and the Indian Ocean,"* edited by N. Van Padang, 1–5. Rome: IAVCEI, 1963.

Workman, W. "The Significance of Volcanism in the Prehistory of Subarctic Northwest North America." In *Volcanic Activity and Human Ecology*, edited by P. Sheets and D. Grayson, 339–71. New York: Academic Press, 1979.

Wood, B. "Did the Israelites Conquer Jericho?" *BAR* 16 (March–April 1990): 45–58.

Woods, A., and K. Wohletz. "Dimensions and Dynamics of Co-ignimbrite Eruption Columns." *Nature* 350 (1991): 225–27.

Yadin, Y. "Is the Biblical Account of the Israelite Conquest of Canaan Historically Reliable?" *BAR* 8 (March–April 1982): 16–23.

Yakir, D., S. Lev-Yadun, and A. Zangvil. "El Nino and Tree Growth near Jerusalem over the Last 20 Years." *Global Change Biology* 2 (1996): 97–101.

Yalciner, A., E. Pelinovsky, A. Zaitsev, A. Kurkin, C. Ozer, H. Karakus, and G. Ozyurt. "Modeling and Visualization of Tsunamis: Mediterranean Examples." In *Tsunami and Nonlinear Waves*, edited by A. Kundu, 273–83. Berlin: Springer, 2007.

Yurco, F. "Merenptah's Canaanite Campaign," *Journal of the American Research Center in Egypt* 23 (1986): 189–215.

Yurco, F. "Merenptah's Canaanite Campaign and Egyptian Origins." *Exodus*, 43–55.

Zdanowicz, C., G. Zielinski, and M. Germani. "Mount Mazama Eruption: Calendrical Age Verified and Atmospheric Impact Assessed." *Geology* 27 (1999): 621–24.

Zertal, A. "The Trek of the Tribes as they Settled in Canaan." *BAR* 17 (September–October 1991): 48–49, 75.

Zertal, A. "'To the Land of the Perizzites and the Giants': On the Israelite Settlement of the Hill country of Manasseh." *FNM*, 47–69.

Zertel, A. "Philistine Kin Found in Israel." *BAR* 28 (May–June 2002): 18–31, 60–71.

Zevit, Z. "Archaeological and Literary Stratigraphy in Joshua 7–8." *BASOR* 251 (1983): 23–35.

Zielinski, G. , P. Mayewski, L. Meeker, S. Whitlow, M. Twickler, M. Morrison, D. Meese et al. "Record of Volcanism Since 7000 B.C. from the GISP2 Greenland Ice Core and Implications for the Volcano-Climate System." *Science* 264 (1994): 948–52.

Index

cinder cones, 54, 59, 60, 63–64, 178n50
circumcision, 72
climatic conditions, 11, 13, 82–85, 117, 145,
  150, 163n7; in East Africa, 84–85, 145,
  182nn9 and 10; in Europe, 83–84, 86, 145,
  182n5; in the Middle East, 83, 86, 188n35;
  in the Sahara, 84, 86, 145
Coats, George, 36
co-ignimbrite eruption column. See eruption
  column
Conquest, biblical narrative of, xiv, xv, 2, 3, 9,
  107–109, 113–14, 148, 149, 150. See also
  Canaan, settlement of by Israelites
cooking pots, 18–19, 164n35, 175n6, 184n42,
  197n21
counterdisaster syndrome, 33, 45
covenant renewal sacrifice, 61, 74, 79, 99, 128,
  138, 141, 142, 150. See also Passover
Crater of the Full Moon. See Hala'-l-Bedr
craters, 54
Crete, 27, 29, 135
crisis cult. See nativistic movements
Cross, Frank Moore, xviii, 116, 140
cultic shrines, 116, 117. See also Lachish;
  Shechem; Shiloh
cups, 19
Cushan, 81, 116
Cushan-rishathaim of Aram-naharaim,
  115–16
Cushite, 71, 81, 116
cyclonic storms, 28, 41
cylinder seal, 13, 50
Cyprus, 21, 47

Dan, 102, 114, 115
Daphnae, 48
darkness, xiii, 6, 41, 126, 135–36, 148
Dathan, 70, 71
David, 4, 69, 115, 116
Davies, Graham, 66, 177n40
Dead Sea, 80, 81, 98, 112–13, 116, 118, 145,
  150, 188n35
Dead Sea pull-apart basin, 82, 95
Dead Sea Rift, 54, 99, 102
Debir, 108, 115
Deborah, 114, 140; Song of, 81–82, 197n22
Decalogue, 63, 73
Deir el-Bahri, 137
Deir el-Medinah, 127

Delta of the Nile, 1, 2, 6, 9, 10–13, 16, 27, 28,
  33, 36, 38, 40, 41, 44, 48, 51, 87, 89, 123,
  124, 126, 128, 129, 135, 141, 143, 148
Demetrius, 56
DeMille, Cecil, 1
dendrochronology. See tree-ring chronologies
Deukalion's flood, 133, 134, 135, 194n52
de Vaux, Roland, 36, 42
Dever, William, 2, 3, 13, 15, 21, 164n35
Dhiban, 57, 82
Documentary Hypothesis, 35, 62, 177n43.
  See also Elohist; Priestly source; Yahwist
donkeys: caravans and caravanners of, 14, 19,
  20, 37, 52, 68, 91, 174n44, 184n42; sacrifices
  of, 13, 20, 42, 101; at Tell el-Maskhuta, 18
Doughty, Charles, 64, 74, 80, 178n50
drought, 145
Dynasties: Eighteenth, 1, 2, 8, 9, 24, 43, 84, 88,
  120, 124, 125, 127, 138, 192n31; Fifteenth
  (Hyksos), 13, 15, 16, 20, 51, 88, 164n16;
  Fourteenth, 13, 14, 47, 48, 49, 50, 175n3;
  Nineteenth, 2, 3, 22, 84, 124, 127, 137, 138,
  143–44; Seventeenth, 16, 87, 88, 137;
  Sixteenth, 175n3; Thirteenth, 13, 14, 20, 49,
  50, 163n7, 175nn3 and 6, 192n21; Twelfth,
  11, 13, 163n7, 174n44; Twentieth, 2, 3;
  Twenty-First, 137, 138; Twenty-Sixth, 2

Early Bronze Age, 105
earthquakes, xix, 72, 95, 97, 99, 100–101, 111,
  134, 147, 150; with volcanic eruptions, 61,
  62. See also seismic waves
East Africa, 69, 86, 87
Ebenezer, 147–48
Ebola virus, 84, 183n19
Edom, 80–82, 116, 118
Edomites, 118, 119
Eglon, 107, 108
Egyptian administrative centers in Canaan,
  196n2
Egyptian military campaigns, 87–90, 110,
  120–24; 145–46, 147
Egyptian religious beliefs, 8, 43, 127, 136, 137,
  195n66
El-Arish naos, 8, 125–26, 127, 128, 129, 130,
  135, 136, 150, 162n38
Eldaah, 52
elders of Israel, 42, 43, 63, 74, 149
Eleazar, 48, 71, 78

Hamor, 109
Hanoch, 52
Har Karkom, 178n46
Har Maron, 102
Harrat al 'Uwairidh, 53, 58, 176n21
Harrat ar Raha, 52, 53, 58, 59, 66
Harris, James E., 137, 138
harvest times: Canaanite, 60–61, 98, 99–100,
    141, 143; in the Egyptian Delta, 5, 6, 36, 60,
    61, 89, 141. *See also* barley; flax; wheat
Hatshepsut, 8, 9, 88, 120, 121, 127
Hattusili I, 181n2
Hawaii, 54
Hawaiite, 176n21
Hayes, Christine E., 77, 180nn29 and 34
Hazor, 108, 144
hearsay, 152
Hebrews, 1, 43, 46, 51
Hebron, 69, 104, 107, 108 109, 115, 116, 138
Hecataeus of Abdera, 87
Hegaz, 67, 79, 184n42
Heliopolis, 127, 175n5
Hendrix, Ralph E., 77
Herculaneum, 26
Herzog, Chaim, 106, 107
Hesiod, xviii, 33
Hesron, 116
Heston, Charlton, 1
hieroglyphics, 14, 80
historical gossip, xviii
Hittites, 69, 83, 145
Hobab, 75, 76, 115
Hoffmeier, James K., 130
Holladay, John S., Jr., 17, 19, 20, 22
Homer, xvii, xviii
homogenites, 26
Horeb, 7, 51, 58, 66, 78
Hormah, 115
hornets, 90, 185n43
Horn of Africa, 19, 30, 86
horses, 13, 15, 31, 122, 129, 130, 136
Hort, Greta, 6–7, 44, 52, 67, 70, 75
Horus, Way of, 122, 123, 129, 138, 143
Hoshea, son of Nun, 80
human sacrifice, 14, 32
Hur, 73, 75–76, 150
Hurrians, 69, 83, 111, 112
Hyksos, 2, 3, 4, 9, 10–16, 19–22, 33–34, 37, 42,
    44, 51, 55, 69, 80, 87–91, 110, 122, 125, 127,

141, 149, 150, 192n31. *See also* Dynasties,
    Fifteenth

Ice Age, 16, 95
ice cores, 23, 24, 165n2, 166n6
ice in eruption clouds, 30, 41
ignimbrites, 26
implicational errors, xvi, 39, 153, 155
Inanna, 38
insects, 41
Intertropical Convergence Zone, 13, 84
Iron Age, xiv, 105, 106, 111, 115, 144
iron, in seawater, dust, or tephra, 38–39,
    172n20
Ishmael, 49
Isis, 137, 195n66
Israel: on the Merneptah stela, xv, 145–46; oral
    traditions of. *See* oral traditions, Israelite;
    tribal boundaries of, 113–14, 116, 150;
    tribal league of, xviii, 148; tribes of, 112–16,
    118; village settlement of, 144–47
Israelite fertility, 190n69
Israelite-Gibeonite alliance, 107–108
Israelite monarchy, xviii, 148, 151
Issachar, 114
ivory, 19, 68–69, 86–87, 89
Iye-abarim, 82

J source. *See* Yahwist
Jabon, 109
Jack, J. W., 4, 161n16
Jacob, 16, 46, 48, 51, 74, 109, 116
Jarmuth, 107, 108
Jebusites, 69
Jephthah, 118
Jephunneh, 75
Jerahmeelites, 116, 119
Jericho, xiv, 4, 9, 69, 82, 92, 93–101, 104, 107,
    141, 149, 150, 186n12
Jericho East Fault, 94, 95, 97
Jerusalem, 104, 105, 107, 108, 109, 112, 116,
    144
Jethro, xiii, 76
Jews, 89, 125, 175n5
Jezreel Valley, 103, 104
Jochebed, 48, 49, 50, 192n21
Jordan, kingdom of, 58, 81
Jordan River, xiv, 52, 81, 82, 92, 98, 99, 102,
    104, 113, 116, 150

rats, 85, 86, 90, 184n39

Re, 8, 136. *See also* Amun-Re

Rebekah, 48, 50

Redford, Donald, 2–3, 8, 87

Redmount, Carol A., 17, 20, 21, 22

Red Sea, 7, 12, 19, 52, 57, 58, 86, 95, 132, 136, 143, 174nn44, 45, and 46

red tides, 38–39

Reed Sea, 130, 132, 138, 141, 143, 151

reindeer, 31, 40, 170n50

Rekhmire, 121, 173n42, 190n2

Rephidim, 71, 77

Retame, 66

retrieval errors, xvii, 153

Reuben, 82, 91, 113, 114, 116, 119, 150

Reubenite revolt, 70–71, 79, 179n13

Reuel, 75

Revisionists, xiv

revitalization movement. *See* nativistic movements

Richter scale, 99

Rift Valley volcanoes, 7

Rithmah, 66

ritual numbers, 5, 64, 126. *See also* forty years

Roman Britain, 111–112

royal scepter (Degai), 126, 129

ruins of Ai, 105–107

rumor, 152, 154–55; leveling of, 40, 141, 154, 155, 175n5; sharpening of, 37, 154; structuring of, xvii, xviii, 154–55

Saft el-Henna *naos*. See El-Arish *naos*

Saite period, 2, 22

Salitis, 11, 16, 51, 88. *See also* Sheshi

Sara, 48, 50

Sargon the Great of Akkad, 49

Saul of Tarsus, 56, 58

scarabs, 14, 15, 16, 18, 20, 21, 50, 51, 69, 122

Sea Peoples, 147. *See also* Philistines

secondary maxima, 30, 41

Second Intermediate Period, 11, 12, 15, 16, 17, 24, 86–87, 123

Seir, 81–82, 116; Way of, 56, 57

seismic waves, 100–101

Sellin, Ernst, 93, 94, 97, 98

sêneh, 65

Seqenenre Tao, 87

Sesostris, 174n44

serial recall, 125. *See also* rumor

Seth, 14, 33, 37, 42, 195n66

Sethnakht, 3

Seti I, 173n42, 174n44, 195n73

Seti II, 138

Sharuhen, 68, 89, 110, 111, 120. *See also* Tell el-'Ajjul

shasu, 114, 118–19, 120, 129, 132, 143, 146, 150

Shea, William, 124

Shechem, 104, 109, 117, 118

sheep, xiii, 13, 22, 31, 40, 50, 127; sacrifices of, xiv, 43, 101, 173n38; revered by Egyptians, 43, 127

Shephelah, 102, 107, 108

Sheshi, 16, 51, 69, 163n13

Sheth, 81

Shi-Hor, 122, 129, 136, 193nn35 and 37

Shiloh, 4, 117, 148

Shiprah, 46

Shittim, 82, 91, 98, 116

Shu, 126

Shur, Way of, 57, 138

Shutu, 81

Sile. *See* Tell Hebua I

silos, 14, 18, 146

Simeon, 109, 115, 116

Simeonites, 116, 119

Sin, 65

Sinai Peninsula, 16, 19, 45, 51, 52, 65, 87, 89, 120, 123–24, 143, 144, 174n44

Sinai, revelation and covenant at, 62–63, 64, 73, 140, 143

Sinuhe, 81

Sirbonis lagoon, 7

slaves, 6, 46, 119, 121–24, 129, 141, 144, 150

slavery. *See* slaves

Sobekhotep III, 46

Sobekhotep VII, 15

soils, 102

Sojourn in the Wilderness, xiv, xv, 70, 72–73, 148; itinerary of, 180n30

Solomon, 4; temple of, 3–4, 5, 124, 148

Speos Artemidos inscription, 8

spices, 19, 69. *See also* frankincense, myrrh

spies, 68, 69, 97–98, 105

Spring equinox, 60, 61, 128, 141, 149–50

Spring of Elisha, 95

Stager, Lawrence, 112

stone quarries, 127

structuring. *See* rumor

Subboreal climate phase, 145

subduction, 25

subduction-zone volcano, 25, 54

Suez, 45; Gulf of, 45, 53, 174n46; isthmus of, 143

Succoth, 12, 17, 44, 45

sulfur, 30–31, 170n44. *See also* acid

sunspot minimums, 84

Syncellius, 125

Syria, 11, 13, 89, 90, 95, 120, 121, 122, 123, 127, 145; princes of, 120

tabernacle, 67, 79

tablets of the covenant, 77, 79–80, 150

Tadra, 58, 177n30

Talmai, 69

Tanis, 1, 175n5

Tati, Nubian princess, 14

Tebûk, 53, 58, 59

tectonic oscillation, 111

tectonic plates, 25, 52, 95

tectonic rift, 95

Tefnut, 126

Tekoa, 115

Tell Defenneh, 129

Tell el-'Ajjul, 21, 68

Tell el-Borg, 123, 129, 130

Tell el-Dab'a, 11–16, 18, 19, 20, 122, 124, 127, 173n38; mass graves at, 14; Tell A section of, 11, 13, 14

Tell el-Maskhuta, 12, 17–22, 37, 44, 50, 51

Tell el-Yehudiyah, 16, 20, 37

Tell er-Retabah, 16, 17, 19, 22, 37

Tell Hebua I, 27, 87, 122, 129, 136

Tell Rumeida. *See* Hebron

Tel Michal, 27, 134, 168n22

telescoping, xvi, 73, 154, 155–56

Ten Commandments: in the Bible, 79; movie of, 1

tephra, 25, 27–32, 39, 41, 135–36, 150, 168n28. *See also* ash; volcanic glass

Tethmosis, 125. *See also* Tuthmosis III

Thebes, 1, 8, 16, 22, 47, 87, 127

Theogony, The, 33

theophany, 52, 63

Thera. *See* Minoan eruption; volcanoes, Santorini

Therasia, 26

thermoluminescence dating, 134, 202–203

Thum, 126

Thirteenth Dynasty rulers, 15. *See also* Dynasties, Thirteenth; Khaneferre Sobekhotep IV; Neferhotep I, Sobekhotep III

Thummin, 78

Thummosis. *See* Tuthmosis III

thunderstorms, 62

Tilos, 135

Timsah, 16, 45, 173n44

Tjaru. *See* Tell Hebua I

tombs: at Tell el-Maskhuta, 18, 20, 50, 51, 69; at Jericho, 20, 90

Torah, 143

toxic dinoflagellates, 38

trade, 14, 19, 22, 68, 86–87, 90, 184n42

Transjordan plateau, 104

transposition. *See* retrieval errors

tree-ring chronologies, 23; absolute, 24; Anatolian, 23, 24, 165n2; European, 24, 84; Northern Hemisphere, 28

Trichodesmium, 38, 39, 172n20

tsunamis, xix, 8, 26, 27, 29, 33, 38, 39, 133–36, 137, 148, 150–51, 168n22, 172n13, 193n35

Tupi-Guarani, 55

turbidity, 31, 38

Turin king list. *See* papyri, Turin

Turkey, 27, 95

Tur Sina, 65–66

Tuthmosis I, 8, 88, 89, 120, 137, 138

Tuthmosis II, 88, 89, 120, 137, 138

Tuthmosis III, 8, 82, 88, 89, 120–22, 124–25, 129, 133, 136–38, 141, 150–51, 175n5, 190n2. *See also* Pharaoh of the biblical Exodus

Tuthmosis IV, 137, 138

typhus, 14

Ugarit, 40, 145

Ullikummi, Hittite myth, 33, 171n62

Upper Egypt, 19, 22, 175n6

Uri, 73

Urim, 78

Valley of the Kings, 137

Valley of Trouble, 116

Van Seters, John, 35–36

Vansina, Jan, xvi, 140, 155

Vitaliano, Dorothy, 7, 174n46

Volcanic Explosivity Index (V.E.I.), 25, 162n29

volcanic glass, 23, 27, 30, 165n2, 168n28

volcanoes: Aniakchak, 23, 24, 83, 165n2; Huaynaputina, 28–30, 32; Ilopango, 33; Karkar, 31; Katmai-Novarupta, 32, 40; Kilauea, 54; Krakatoa, 25, 26, 30, 86; Long Island, New Guinea, 32, 40; Mayon, 32; Paricutin, 31, 32; Santorini, xix, 23–27, 28, 30, 54; Tambora, 25; Unimak, 40, 170n50; Vesuvius, 25, 26; White River, 33. *See also* Minoan eruption; Mount Etna; Mount Spurr; Mount St. Helens

volcanic eruptions, xix, 23, 61–62, 134, 148; climatic effects of, 83; effusive, 7; explosive, 25; plinian, 25–26, 27–28, 36. *See also* Minoan eruption

Voltzinger, N., 174n46

von Wissmann, 59

Wadi al-'Arabah, 72, 75, 76, 80, 82, 115

Wadi Dabr, 116

Wadi el-Hesa. *See* Wadi Zered

Wadi el-Mūjub, 82

Wadi Tumilat, 16–22, 34, 37, 42, 43, 45, 50, 51, 68, 70, 72, 124, 141, 149, 174n44

Wadi Zered, 72, 82, 118

walls at Jericho: Early Bronze Age, 93, 95; lack of in Late Bronze Age, 93; Middle Bronze Age, 93, 94, 96, 98, 101, 186n25

Ward, William, 3, 9

Waterloo, battle of, xv–xvi

Watzinger, Carl, 93, 94, 97, 98

weaning, 56, 190n69

weapons, 11, 18, 20, 69, 132, 136

weather systems, 28. *See also* cyclonic storms

Weinstein, James, 3, 20, 21, 22

Wells of Merneptah, 146

Wente, Edward F., 137

wheat, 18, 32, 60, 100

wilderness, xiv, 2, 56, 72, 150; of Sin, 65; of Sinai, 60, 66. *See also* Sojourn in the Wilderness

Wilson, Ian, 7, 172n12

winds, 25, 27–28, 41, 45, 132, 163n7, 174n46

Xenopsylla cheopis, 86, 90

X-ray analysis of mummies, 137–38

Ya'ammu, 15

Yahweh, xiii, 48, 61, 118

Yahwist (J source), 35, 36, 63, 130

Yakbim, 15

Y'akub-Hr, 16

Yakut, 65

Yali, 134, 135, 140, 141, 142, 148, 150

Yano'am, 145

Yanomami, 117

Ya'ush (Ye'ush), 81

year counts, 4–5, 148; by Maricopa and Lakota Indians, 161n20

YHWH, 63, 135, 138

Yurco, Frank, 1–2

Zebulun, 114

Zevit, Ziony, 105

Zimri-Lim, 92, 101

Zipporah, 72